Physical activity and exercise affect many dimensions of health. This book presents an up-to-date and wide-ranging account of the key issues of the biology of physical activity and health. The context is set by considering the comparative and temporal aspects of activity in humans. There follows an examination of the concepts and methodological issues associated with activity, exercise, health and fitness as well as their interrelationships. Particular attention is given to activity in children, adolescents and the elderly, and the psychological effects of activity. The book ends with an overview of current and future leisure lifestyles.

This topical volume will be of interest to human biologists, biological anthropologists, human physiologists, sports scientists, psychologists, and healthcare professionals involved in preventative medicine, public health, geriatrics, occupational health and community medicine.

SOCIETY FOR THE STUDY OF HUMAN BIOLOGY SYMPOSIUM SERIES: 34

Physical activity and health

PUBLISHED SYMPOSIA OF THE SOCIETY FOR THE STUDY OF HUMAN BIOLOGY

10 Biological Aspects of Demography ED. W. BRASS
11 Human Evolution ED. M.H. DAY
12 Genetic Variation in Britain ED. D.F. ROBERTS AND E. SUNDERLAND
13 Human Variation and Natural Selection ED. D.F. ROBERTS *(Penrose Memorial Volume reprint)*
14 Chromosome Variation in Human Evolution ED. A.J. BOYCE
15 Biology of Human Foetal Growth ED. D.F. ROBERTS
16 Human Ecology in the Tropics ED. J.P. GARLICK AND R.W.J. KEAY
17 Physiological Variation and its Genetic Base ED. J.S. WEINER
18 Human Behaviour and Adaptation ED. N.J. BLURTON JONES AND V. REYNOLDS
19 Demographic Patterns in Developed Societies ED. R.W. HIORNS
20 Disease and Urbanisation ED. E.J. CLEGG AND J.P. GARLICK
21 Aspects of Human Evolution ED. C.B. STRINGER
22 Energy and Effort ED. G.A. HARRISON
23 Migration and Mobility ED. A.J. BOYCE
24 Sexual Dimorphism ED. F. NEWCOMBE *et al.*
25 The Biology of Human Ageing ED. A.H. BITTLES AND K.J. COLLINS
26 Capacity for Work in the Tropics ED. K.J. COLLINS AND D.F. ROBERTS
27 Genetic Variation and its Maintenance ED. D.F. ROBERTS AND G.F. DE STEFANO
28 Human Mating Patterns ED. C.G.N. MASCIE-TAYLOR AND A.J. BOYCE
29 The Physiology of Human Growth ED. J.M. TANNER AND M.A. PREECE
30 Diet and Disease ED. G.A. HARRISON AND J. WATERLOW
31 Fertility and Resources ED. J. LANDERS AND V. REYNOLDS
32 Urban Ecology and Health in the Third World ED. L.M. SCHELL, M.T. SMITH AND A. BILSBOROUGH
33 Isolation, Migration and Health ED. D.F. ROBERTS, N. FUJIKI AND F. TORIZUKA

Numbers 1–9 were published by Pergamon Press, Headington Hill Hall, Headington, Oxford OX3 0BY. Numbers 10–24 were published by Taylor & Francis Ltd, 10–14 Macklin Street, London WC2B 5NF. Further details and prices of back-list numbers are available from the Secretary of the Society for the Study of Human Biology.

Physical Activity and Health

34th Symposium Volume of the
Society for the Study of Human Biology

EDITED BY

N.G. NORGAN
Department of Human Sciences
Loughborough University of Technology

CAMBRIDGE UNIVERSITY PRESS
Cambridge, New York, Melbourne, Madrid, Cape Town, Singapore, São Paulo

Cambridge University Press
The Edinburgh Building, Cambridge CB2 8RU, UK

Published in the United States of America by Cambridge University Press, New York

www.cambridge.org
Information on this title: www.cambridge.org/9780521415514

First published 1992
This digitally printed version 2008

A catalogue record for this publication is available from the British Library

Library of Congress Cataloguing in Publication data

Society for the Study of Human Biology. Symposium (34th : 1992 : Oxford University) Physical activity and health: 34th Symposium volume of the Society for the Study of Human Biology / edited by N.G. Norgan.
p. cm. – (Society for the Study of Human Biology symposium series ; 34)
Includes index.
ISBN 0 521 41551 9 (hardback)
1. Exercise–Physiological aspects–Congresses. 2. Physical fitness–Congresses. 3. Health–Congresses. I. Norgan, N.G. II. Title. III. Series.
[DNLM: 1. Exertion–congresses. 2. Health–congresses. 3. Physical Fitness–congresses. W1 SO861 v.34]
QP301.S75 1992
613.7′1–dc20 92–12278 CIP
DNLM/DLC
for Library of Congress

ISBN 978-0-521-41551-4 hardback
ISBN 978-0-521-06746-1 paperback

Contents

List of contributors

Dr N.G. Norgan
Department of Human Sciences, University of Technology, Loughborough, Leics. LE11 3TU, UK

Professor R. McNeill Alexander, FRS
Department of Pure and Applied Biology, University of Leeds, Leeds LS2 9JT, UK

Professor J.V.G.A. Durnin
Institute of Physiology, University of Glasgow, Glasgow G12 8QQ, UK

Dr J.R.Kemm
Department of Social Medicine, University of Birmingham, Birmingham B15 2TT, UK

Dr P.S.W. Davies
MRC Dunn Nutritional Laboratory, Downhams Lane, Milton Road, Cambridge CB4 1XJ, UK

Dr C.B. Cooke
Activity & Health Research, School of Sports and Exercise Science, University of Birmingham, PO Box 363, Birmingham B15 2TT, UK

Professor H.G.C. Kemper
Department of Health Science, Faculty of Human Movement Sciences, Vrije Universiteit & Medical Faculty, Universiteit van Amsterdam, Meibergdref 15, 1105 AZ Amsterdam, The Netherlands

Professor R.M. Malina
Department of Kinesiology & Health Education, University of Texas at Austin, Texas 78712, USA

Dr E.J. Bassey
Department of Physiology & Pharmacology, Medical School, Queens Medical Centre, Nottingham NG7 2UH, UK

Dr K. Morgan
Department of Health Care of the Elderly, Medical School, Queens Medical Centre, Nottingham NG7 2UH, UK

Dr A.E. Hardman
Department of Physical Education & Sports Science, University of Technology, Loughborough LE11 3TU, UK

Professor W.H.M. Saris
Department of Human Biology, University of Limburg, 6200 MD Maastrict, The Netherlands

Dr S. Biddle
School of Education, University of Exeter, St Luke's, Heavitree Road, Exeter EX1 2LU, UK

Dr C. Panter-Brick
Department of Anthropology, University of Durham, 43 Old Elvet, Durham DH1 3HN, UK

Professor A. Steptoe
Department of Psychology, St George's Hospital Medical School, Cranmer Terrace, London SW17 0RE, UK

Professor S. Glyptis
Department of Physical Education & Sports Science, University of Technology, Loughborough LE11 3TU, UK

Acknowledgements

The support of the following organizations towards the cost of the Symposium is acknowledged: The British Council, British Heart Foundation, The Royal Society and the Sports Council.

Symposium sessions were chaired by Professor J.C. Waterlow, FRS, Professor C. Williams, Professor R. Goldsmith and Professor E.J. Clegg. Their contributions and the work of the staff of the Department of Human Sciences, Loughborough University, is gratefully acknowledged.

1 *Introduction*

N.G. NORGAN

It is surprising how few attempts have been made to describe or to reach a consensus on current knowledge on physical activity, exercise and health. The 32 published Symposia of the Society for the Study of Human Biology, for example, do not include a title on this topic. This is unexpected for a Society now nearly 35 years old and containing many members interested in this area. Recent symposia have examined Energy and Effort (Harrison, 1982) and Capacity for Work in the Tropics (Collins & Roberts, 1988), and disease and abnormality have been a recurring interest in the Society's symposia but physical activity and health has been left to others.

It might be thought that there was no need for another symposium on the topic. The case for exercise has been made (Fentem and Bassey, 1978) and updated (Fentem, Bassey & Turnbull, 1988) and the leading experts met in Toronto in 1988 to produce a 700-page consensus of current knowledge on exercise, fitness and health (Bouchard *et al.*, 1990). In the United Kingdom, sport, health, psychology and exercise were examined with a mental health emphasis at a three-day symposium in 1988 emanating from the Fitness and Health Advisory Group of the Sports Council and the Health Education Authority (*Proceedings of Symposium on Sport, Health, Psychology and Exercise*) (n.d.). What these latter sources disclose is that there is still much to be understood. Understanding and concepts of health are being revised, models describing the complex relationships between habitual physical activity and health are becoming more sophisticated and methods and study designs are improving although some critical research issues remain.

Physical activity has not proved amenable to measurement. The lack of suitable techniques has slowed our understanding of the role of physical activity and exercise in promoting health. It has also contributed to false and conflicting claims of the importance and efficacy of physical activity. Over the years, various instruments and procedures to assess physical activity have been developed, usually with high initial expectations that have not been met. Current interest is in the doubly labelled water method for measuring total daily energy expenditure. The method is technically difficult, but high capital and running cost may limit its operation to competent laboratories and the method should provide valuable information on the energetics of activity and exercise. So far, the data base is small and the representativeness of the subjects becomes a key issue demanding attention.

The United Kingdom has lagged behind many other countries in not having representative data on physical activity and fitness of its inhabitants. The Government has recently published a consultative document on health in England, 'The health of the nation' (Department of Health, 1991) with the objective of identifying the main health problems and focussing as much on the promotion of good health and the prevention of disease as on the treatment. Key areas were selected on the basis of (i) being a major cause of death, (ii) effective intervention is possible, and (iii) it is possible to set objectives and targets and to monitor progress. In relation to physical activity, however, it is unable to set targets because of the lack of data. This sad, sorry state should be ameliorated by data on current levels of participation in physical activity, exercise and sport and current levels of fitness collected in the Allied Dunbar National Fitness Survey began in 1990 and due to report in 1992. The survey is described in Chapter 6.

The aims of the Symposium and the chapters in this book are therefore two-fold. To hear of, and discuss, the newer studies and what they tell us from a wider perspective of the problem than would be common in a solely physiological based inquiry. Human biologists are interested in the comparative and evolutionary aspects, is man an active or inactive animal?, is modern man noticeably less active than his recent or distant ancestors?, as much as the physiological and psychological. Bringing together these different wide perspectives with emphasis on the information from better designed studies with the newest measurement procedures and instrumentation provides a modern synthesis, leading hopefully to implementation of new findings.

The first two chapters deal with comparative and temporal aspects of physical activity. McNeill Alexander considers man's basal metabolic rate, daily and maximal metabolic rate, energy cost of locomotion and degree of mobility (the home range) in comparison with other animals, allowing for differences in body size. His conclusion is that we are neither remarkably active nor inactive, except in cities where we walk very fast. By most measures of activity, we are more like pigs than the other animals.

Durnin found it difficult to comment on physical activity past and present as so little factual information exists. This situation has given rise to the national surveys of fitness and activity described in Chapter 6. Durnin has found that surveys of children of similar body weight show falls in energy intake of 20% over the last 50 years, confirming the impression of current low levels of activity. The evidence is not unequivocal as Davies in his chapter on developments in the assessment of physical activity presents total daily energy expenditures based on the doubly labelled water technique and expressed as multiples of basal metabolism to be 1.8–2.0, indicative of several hours of hard physical activity. Saris also found values of 1.7–1.8 in the obese and non-obese adolescents using the same techniques. It could be that any secular trend to inactivity has not affected these age-groups as much as is commonly thought.

The question and importance of definitions and validity of measures is considered in several of the chapters. Both Malina and Biddle include clear and apposite definitions of physical activity, exercise and fitness. Physical activity refers to 'any bodily movement produced by skeletal muscles and resulting in energy expenditure'. Exercise is used to refer to structured forms of physical activity usually for reasons of gaining or maintaining fitness. Similarly, appropriate outcome variables are a concern of most investigators and authors. These are considered systematically by Kemm in Chapter 3 on the validity of health measurements. There is no common construct of health. Global measures of health reverse the process of breaking down the concept of health into its constituent dimensions but problems exist in the identification and scaling of the different dimensions. Value judgements are difficult to avoid but, Kemm concludes, these could, and should, be made more explicit to aid in the interpretation of the data.

The following four chapters by Kemper, Malina, Bassey and Bennett and Morgan examine physical activity and its sequalae in the young and elderly. Kemper and Malina provide reviews of the physical and behavioural development in children and youths, including much of their own work while Bassey and Morgan concentrate on the Nottingham Longitudinal Study of Activity and Ageing began in 1985 on a group representative of the United Kingdom national population. Some data on longitudinal changes are now available and are described by Bennett and Morgan.

There is particular concern over the apparent low level of activity in children. There is the argument that, if children do not experience or learn activity patterns or involvement in childhood, they are less likely to be active adults and will suffer the consequences. This is addressed in the chapters by Kemper, Malina and Biddle. Because of the difficulty and dubious ethics of a study randomly assigning sedentary children to activity and non-activity over several years, Kemper has adopted the approach of a longitudinal study of active and inactive teenagers in the Amsterdam Growth and Health Study. Physical activity and aerobic fitness was higher in boys than girls, but fell in both groups over the period 12–18 years. There was no evidence for a fall in fitness over the last 25 years. The active groups appear to be fitter because they are active rather than being more active because they are fitter. Teenage fatness was most closely correlated with coronary heart disease risk factors in young adulthood. These findings illustrate and emphasize the importance of preventative strategies at an early age. This conclusion is echoed by Malina who calls for the systematic study of physical activity to become a major objective in human biology research. Activity influences the domains of growth, maturation and development and, as illustrated by Malina, the development of competence in a variety of behavioural domains is, 'one of the major tasks of childhood and youth'.

Considerable attention has been focussed on activity and health in the elderly too. The expectation is that inactivity will reduce the capacity for activity and fitness, leading to dependence and possible institutionalization. In the elderly, the concepts of quality of life supersede those of length of life. The retired and elderly are seen as appropriate candidates for health promotion initiatives but, as regards mental health particularly, relatively little research has directly addressed the assumption that customary physical activity contributes to psychological well-being in later life. The Nottingham Longitudinal Study of Activity and Ageing described in the chapters by Bassey and Bennett and Morgan was set up to assess the role of lifestyle, including customary physical activity in promoting mental health and psychological well-being in later life. Longitudinal changes in activity are characterized by drift rather than by shift while morale is stable rather than changing. In some activities, however, there is clear evidence of an activity–effect relationship.

Hardman, too, touches upon the health of the elderly, specifically on the reduction of functional capacity and the problem of osteoporosis, in her account of the potential of modest amounts of physical activity to promote health. This can operate by maintaining or increasing functional capacity, influencing energy balance, decreasing the likelihood of disease and in contributing to the management of patients with existing diseases. There is convincing evidence for the first outcome, and sufficient evidence for the other three, but Hardman identifies the problem of the prescription of appropriate activity for individual objectives and circumstances. Low intensity exercise such as brisk walking may meet the needs of many individuals and circumstances.

Saris considers more fully the role of activity in obesity and weight maintenance. Obesity is a major health problem in western society and increasing energy expenditure might appear an obvious way of reducing the excess energy stores. Saris describes the methodological difficulties in measuring energy intake, expenditure and stores and finds conflicting evidence over the role of activity in obesity. It is often thought that engaging in exercise may result in less physical activity at other times of the day. New data considered here suggest this is not the case. Indeed, resting metabolism may be increased after exercise and in the trained state. The gender differences in the effect of activity on energy balance are interesting and of potential importance. Overweight and obesity may be related to the capacity to oxidize fat. Given the regional differences in fat distribution in the sexes and the variations in muscle fibre type with their different metabolic characteristics, physical activity may have a much greater influence on body fat stores in individuals or the sexes than simply the amount of energy expended.

Although most of us and the rest of the population accept that exercise is good for us, involvement in physical activity and exercise is at a low level. It is a research priority, therefore, to identify those factors influencing the adoption and maintenance of physical activity and exercise. Biddle's chapter is a detailed summary of current knowledge on the likely psychological correlates and proposed determinants of participation in activity and exercise by children and adults. For children, establishing positive feelings about activity is important if involvement as adults is to be achieved. Spontaneous intermittent bouts of activity influenced by fashion and season are more appropriate than prolonged periods of activity.

Women in many parts of the world often have little choice over adhering or not to physical activity patterns. Survival and reproductive success may depend on maintaining activity levels including throughout much of pregnancy. Panter-Brick's chapter explores the relationships between women's work and maternal–child health in rural Nepal. Women manage to sustain levels of physical activity despite ill-health and a potential conflict between economic and child bearing responsibilities by a combination of behavioural strategies such as the flexibility of labour exchange and the energetically efficient organization of tasks as well as physiological and mechanical adaptations resulting from the physical training of early childhood and the pace of endurance work.

Steptoe proposes three major reasons for considering the effects of physical activity on mood and mental health; that mental health is an important component of well-being; that activity may help people cope with stress more effectively, and that an understanding of the beneficial psychological consequences may help to enhance adherence to training or rehabilitation programmes. He shows that there is now sufficient evidence to conclude that physical activity does have positive long-term benefits on mood and psychological well-being. The exercise may be aerobic and anaerobic but should not be too light or too heavy. The mechanism is less clear. The effect is more immediate than and independent of the changes in fitness. There are promising avenues for future work in this area.

It is appropriate that the last contribution includes a look to the future. Glyptis gives a view of leisure lifestyles in the present and future. Leisure is taken to mean any activity (or inactivity) freely undertaken for the sake of enjoyment. Thus, it depends on the meaning of the activity to the individual, toil to one person may be leisure to another. Predictions of 20–25 years ago that people were fast moving to an age of leisure have only been fulfilled partially, and sport and exercise participation are still relatively low. Glyptis regards the most significant influences on future leisure patterns to be social polarization and the rise of the 'connoisseur consumer', social fragmentation leading to smaller and more diverse households, and the role of the home as a leisure centre. For most people, leisure is an important component of lifestyle, quality of life and an affirmation of their social position. Leisure alone is rarely fulfilment enough as studies of the unemployment show.

In conclusion, physical activity and exercise affect many of the dimensions of health. It is important that these are identified in order to avoid over- or underestimating the magnitude and significance of the effects of physical activity and exercise. The amount and intensity of activity and exercise needed to attain health benefits may be less than those recommended for fitness benefits. This should further encourage the inclusion of physical activity as a key area in the development of future health strategies. It is hoped that this book will contribute to this process.

References

Bouchard, C., Shephard, R.J., Stephens, T., Sutton, J.R. & McPherson, B.D. (eds) (1990). *Exercise, Fitness and Health: A Consensus of Current Knowledge*. Champaign, Ill: Human Kinetics Books.

Collins, K.J. & Roberts, D.F. (eds) (1988). *Capacity for Work in the Tropics*. Cambridge: Cambridge University Press.

Department of Health (1991). The health of the nation: a consultative document on health in England. *Command 1523*. London: Her Majesty's Stationery Office.

Fentem, P. & Bassey, E.J. (1978). *The Case for Exercise*. London: Sports Council.

Fentem, P., Bassey, E.J. & Turnbull, N.B. (1988). *The New Case for Exercise*. London: Sports Council and Health Education Authority.

Harrison, G.A. (1982). *Energy and Effort*. London: Taylor & Francis.

Proceedings of Symposium on Sport, Health, Psychology and Exercise (n.d.). London: Sports Council.

2 *Comparative aspects of human activity*

R. McNEILL ALEXANDER

Introduction

No animal spends the day sitting at a computer terminal, so should the lives of office workers be regarded as unnaturally inactive? No wild animal goes out running for the sake of its health, so should some people be regarded as unnaturally active? This chapter tries to put human activity in perspective by comparing it with the activity of other mammals.

Typical human body masses are about 50 kg for adult women and 70 kg for men. Other mammals range from 2 g shrews to 6000 kg African elephants (Macdonald, 1984). In making comparisons, differences of size will have to be taken into account.

Quantities such as metabolic rates, speeds of locomotion and distances travelled will be compared, which vary with body size. Comparisons could be limited to animals of similar size to man (for example, leopards, some antelopes and large kangaroos), but if that was done there would be very limited data. It seems better to use data for mammals of all sizes to find out how the measures of activity concerned generally vary with body mass. It will then be possible to predict typical values for mammals of the same mass as man, and compare man with those.

The monotremes (platypus and spiny anteaters) and marsupials (kangaroos, opossums, etc) are in some respects more primitive than the other mammals, which are known as eutherians. Humans are eutherians, and most of the comparisons in this paper are with other eutherians.

Metabolism

First metabolic rates, the rates at which biochemical processes in the body use the available chemical energy of food, will be compared. These can generally be determined by measuring the rates at which oxygen is removed from the air by respiration. More energy is obtained from a gram of fat than from a gram of carbohydrate or protein, but more oxygen is needed to oxidize the fat and (very conveniently for physiologists) metabolism using

1 litre of oxygen releases about 20 000 joules of energy, whatever foodstuff is being oxidized (Blaxter, 1989).

Although the principal concern is with activity, rates of energy use in inactivity will be compared initially. The basal metabolic rate is the rate at which energy is used while resting at a comfortable temperature, not too soon after eating. (Metabolism proceeds faster even when resting, immediately after a meal, because energy is used processing the food.) Blaxter (1989) explains more formally how basal metabolic rate is defined and how it should be measured. It is more easily measured for people, who can be told to lie still, than for animals which can not.

Figure 2.1 shows basal metabolic rates of mammals, plotted against body mass. The rates are expressed in watts (joules per second). A rate of X watts implies that oxygen is being used at a rate of $X/20\,000$ litres per second or 0.18 litres per hour. The scales of the graph are logarithmic so that, for example, the distance from 1 to 10 kg along the mass axis is the same as the distance from 10 to 100 kg. This has two advantages. First, it has made it possible to fit a camel and a dolphin on to the graph without having to crowd the coypu and the sloth (each about 1% of the camel's mass) into the extreme bottom left-hand corner. Secondly, graphs of metabolic rate against mass, plotted in this way, generally approximate quite closely to straight lines.

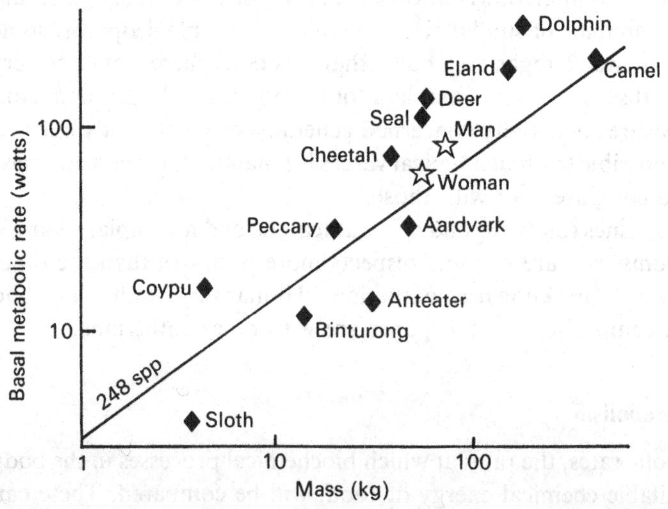

Figure 2.1. A graph on logarithmic coordinates of basic metabolic rate against body mass. The line, from Huyssen & Lacey (1985), is based on data for 248 species of eutherian mammal ranging from a 5 g shrew to a 400 kg camel. The points for humans are from Blaxter (1989), and those for animals have been selected from McNab (1986).

The line shown on the graph is based on data for 248 species, many of them of less than 1 kg mass. Nevertheless, points have been shown on the graph for only a few representative species, ranging from 4 to 400 kg. Notice that mammals of equal mass may have very different metabolic rates. For example, the dolphin has three times the rate predicted by the line and the sloth only half the predicted rate. McNab (1986) showed how these differences could be related to feeding habits. Predators on other vertebrates (such as dolphins, seals and the cheetah) generally have high metabolic rates for their masses, but eaters of invertebrates (anteater and aardvark) have low ones. Herbivores that eat grass and herbs (eland, deer and coypu) have high rates but arboreal mammals that eat fruit and leaves (sloth and binturong) have low ones. These generalizations, illustrated here by just a few examples, are based on McNab's (1986) much more extensive data.

The human metabolic rates shown in Figure 2.1 are probably more truly basal than the others, so should perhaps be increased slightly to make them comparable. Whether such an adjustment is needed or not, they are just a little above the regression line, similar to terrestrial fruit and nut eaters such as the peccary.

Figure 2.2 shows field metabolic rates: i.e. average metabolic rates for normal life. The animal values are averages for a 24 h day and the human ones for a week of five working days and a weekend. They have been obtained by the doubly labelled water technique. In this, the subject is given a small dose of water, in which both the hydrogen and the oxygen are labelled with isotopes. After an interval, the concentrations of the isotopes remaining in the body are measured. The rate of loss of hydrogen isotope gives the rate of loss of water from the body. From this and from the rate of oxygen isotope loss, the rate at which metabolism is producing carbon dioxide can be calculated, because oxygen is lost from the body both in carbon dioxide and in water (Blaxter, 1989).

Field metabolic rates have been measured for many fewer species than basal metabolic rates: the line in Figure 2.2 is based on data for only 23 species. It gives the impression that human field metabolic rates, even of labourers, are low, but this may be an artefact of the restricted sample of species. The only large animals in the data set are three species of sea lions and one of deer. We have already seen that basal metabolic rates of seals and deer are high (Figure 2.1), so their field metabolic rates may also be unusually high, for mammals of their body masses.

The human field metabolic rates are 1.5 times basal for the clerk and twice basal for the labourer. Data in Durnin (1985) show that to increase his rate to that of the labourer, the clerk would have to jog for ten hours per week.

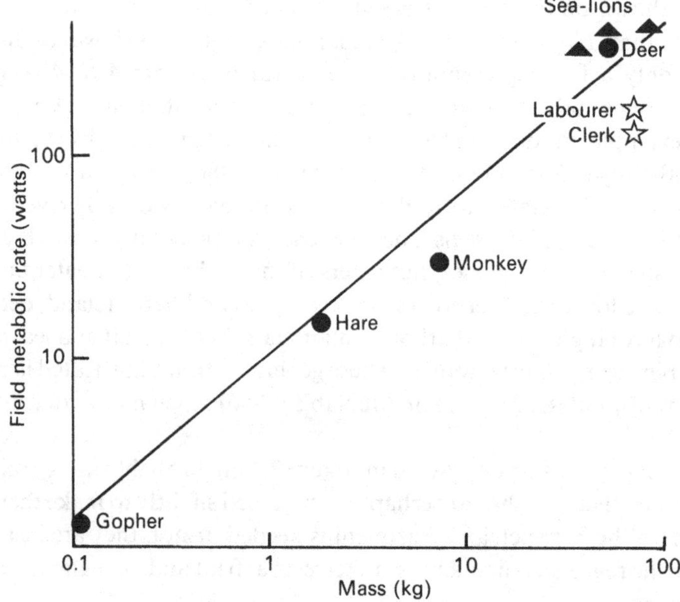

Figure 2.2. A graph on logarithmic coordinates of field metabolic rate against body mass. The line is based on data for 23 species of eutherian mammal, compiled by Nagy (1987). The points for animals are from the same source and those for humans from Durnin (1985).

Figure 2.3 shows metabolic rates corresponding to the maximum rates at which animals can use oxygen in respiration. These are the highest rates of energy use that can be sustained, for example, in distance running. Short bursts of faster energy use are possible, for example, in sprinting, but they depend on processes other than oxidation of food and result in an oxygen debt which cannot be allowed to build up indefinitely (Åstrand & Rodahl, 1986).

Weibel & Taylor (1981) and their associates measured the maximum aerobic metabolic rates of mammals by measuring their rates of oxygen consumption while they ran on treadmills. The animals wore masks connected through flexible tubes to oxygen analysis equipment, so that the air they breathed out could be sucked away and analysed. They were made to run faster by increasing the speed of the tread until no further increase in the rate of oxygen consumption could be obtained. Rates for humans are measured during treadmill running, or pedalling a bicycle ergometer.

The line in Figure 2.3 is based on data for all the 21 species of mammal studied by Weibel & Taylor's (1981) team, some of which had masses of less than 1 kg and so do not appear in our graph. Notice that the points furthest

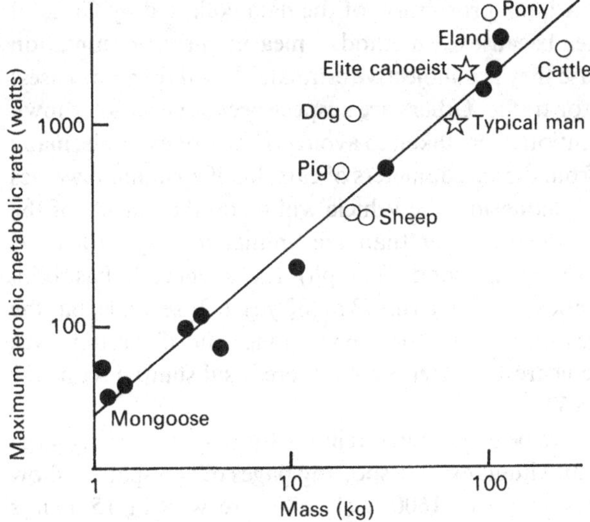

Figure 2.3. A graph on logarithmic coordinates of maximum aerobic metabolic rate, against body mass. The line and the points for animals are from Weibel & Taylor's (1981) study of 21 mammalian species, all but one of them eutherian. The points for humans are from Durnin (1985) and Åstrand & Rodahl (1986). Filled symbols refer to wild animals and open symbols to domestic ones.

above the line are for domestic species (horse and dog) that have been selected for endurance: and that the points for sheep and cattle, domestic species whose endurance seems unimportant, lie well below the line. Human endurance athletes have much higher maximum aerobic metabolic rates than sedentary people: the examples shown are about 20 times the presumed basal rate for élite male canoeists and 12 times basal for more typical young men. Elite swimmers and cyclists have maximum aerobic rates in the same range as canoeists, and even higher rates have been measured for exceptional cross-country skiers. Figure 2.3 shows that human endurance athletes have lower maximum aerobic metabolic rates than would be expected for horses or dogs of their body masses, and that typical people have higher rates than would be expected for cattle or sheep of their masses.

Locomotion

Small mammals such as mice, and large ones such as elephants are relatively slow, but there is no obvious relationship between speed and size for mammals in the range from 10 to 1000 kg (Alexander & Maloiy, 1989). Unfortunately, reliable measurements of maximum speeds are available for very few species. A large proportion of the data collected by Garland (1983) must be rejected because the method of measurement or estimation is unknown, or because they are subjective estimates based (in many cases) on experience of motor traffic. Others are suspect because it is not known whether proper precautions were taken to avoid errors. For example, many depend on readings from the speedometers of cars, but if an animal swerves from a vehicle driving alongside, the vehicle will be on the outside of the curve, so will have to drive faster than the animal to stay with it. A well-known record of 32 m/second (71 mph) for a cheetah has been discredited: it was supposed to have run 73 m (80 yd) in $2\frac{1}{4}$ seconds, but the track on which it was measured is only 65 yd long, and if the time was measured only to the nearest quarter second there is substantial scope for error (Hildebrand, 1959).

Maximum speeds have been measured reliably for only three fairly large species: horses, dogs and humans. The sporting pages of newspapers show that most horse races of up to 1600 m (1 mile) are won at 15–17 m/s (34–38 mph) and most greyhound races at 15–16 m/s. Racehorses and greyhounds have been bred exclusively for speed, so it would not be surprising to find that they were faster than wild mammals which have evolved under natural selection in which other characteristics as well as speed are important. Fast human sprinters reach peak speeds of only 11 m/s (25 mph) (Ballreich & Kuhlow, 1986). Alexander, Langman & Jayes (1977) measured the maximum speeds of mammals ranging from 20 kg gazelles to 1000 kg giraffes, which were pursued by a vehicle in their grassland habitats in Kenya. The methods used were not very accurate but they were at least objective. Maximum speeds of 10–14 m/s were obtained for all species except buffalo, which reached only 7 m/s. These data give the impression that people are rather slow, in comparison with mammals of similar body mass. The impression would be strengthened if some of the very high speed estimates in Garland's (1983) list were accepted.

Elliott *et al.* (1977) made video recordings of lions pursuing their prey, and derived equations showing how speeds increased in the early part of the chase. Their data show that the initial acceleration of the lions was 10 m/s^2 and those of gazelles, wildebeest and zebra about 5 m/s^2 (see Alexander & Maloiy, 1989). Human sprinters accelerate off the blocks at the start of a race at about 10 m/s^2 (Ballreich & Kuhlow, 1986). The performance of the prey species may seem very poor but this may be because they are generally taken by surprise. They might be able to accelerate faster if they could prepare themselves for the start, as lions and human sprinters do.

Weibel & Taylor's (1981) team measured the maximum speeds that mammals can sustain by aerobic metabolism, while running on treadmills. These are, of course, much lower than maximum sprinting speeds. Their data show no obvious dependence of speed on size, in the range from 10 to 250 kg body mass. Sheep had maximum aerobic speeds of 3 m/s and cattle, 4 m/s. Dogs and ponies, however, could sustain speeds that were higher than the maximum speeds of the treadmill. Their maximum aerobic metabolic rates were obtained by tilting the treadmill so that they were in effect running uphill. Metabolic rates were also measured at a range of speeds of level running, making it possible to estimate that the maximum aerobic metabolic rates could have sustained level running at 12 m/s for dogs and 11 m/s for ponies. Maximum aerobic speeds of wild mammals of 10–25 kg mass ranged from 2 to 6 m/s. Men running on treadmills can sustain aerobically speeds up to 6 m/s (Margaria *et al.*, 1963), which is faster than sheep and cattle but very much slower than dogs and ponies. The maximum aerobic speed of humans is very close to that predicted for mammals of our body mass (Garland, Geiser & Baudinette, 1988; data for 39 species of marsupials and placentals ranging from 7 g to 200 kg body mass).

Maximum speeds are used only rarely. Bornstein & Bornstein (1976) measured the speeds of people walking, alone and unencumbered, along the main streets of villages and towns. They found that mean speed depended strongly on the size of the town, ranging from about 0.8 m/s in Greek villages to about 1.7 m/s in cities such as Prague, Munich and Brooklyn. Comparable animal data are rare but Pennycuick (1975) measured mean walking speeds of 0.8 m/s for Thomson's gazelle (mass about 20 kg) and 1.0 m/s for wildebeest (about 200 kg), walking spontaneously in the wild in Tanzania.

Footprints of several hominids that lived 3.7 million years ago have been found at Laetoli, Tanzania. It can be estimated from the size of the feet that they had about the stature of 4–12 year-old children. The lengths of the strides seem to show that they were walking at about 0.5–0.8 m/s (Alexander, 1984).

These data seem to show that human walking speeds in cities are remarkably fast, rather than that village speeds are slow.

It will be asked next, whether human locomotion is economical. Taylor, Heglund & Maloiy (1982) report measurements of oxygen consumption of mammals of a wide range of sizes, running on treadmills at various speeds. For any individual animal, the rate of oxygen consumption is generally found to be related linearly to speed, with the slope of the relationship representing the energy cost of travelling unit distance. This slope may be divided by body mass to obtain the net cost of transport, which is shown in Figure 2.4: this is the energy cost of moving unit mass of animal through unit distance. As a general rule, it falls with increasing body mass, but mammals of equal mass may have very different costs of transport. Most of

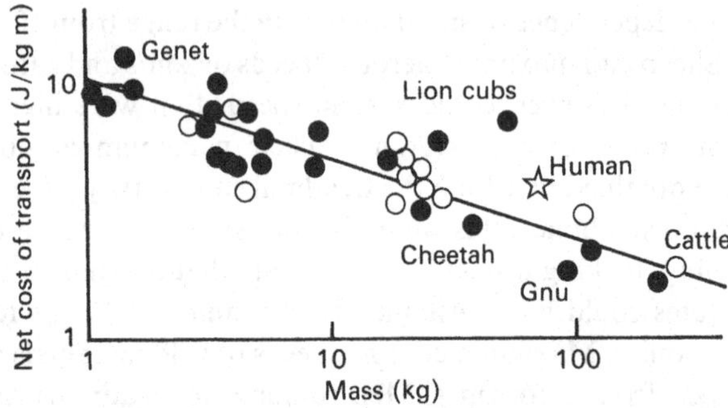

Figure 2.4. A graph on logarithmic coordinates of net cost of transport for terrestrial locomotion, against body mass. The line and all the points are from Taylor, Heglung & Maloiy (1982). Filled symbols refer to wild and open symbols to domestic mammals. The data set on which the line is based refers to 63 species.

the deviations from the regression line are difficult to explain. The graph gives the impression that human locomotion is rather expensive, but the point refers to running. A point has not been added for human walking because, for it, the relationship between oxygen consumption and speed is so far from being linear that it seems impossible to define a net cost of transport (Margaria, 1976). However, rates of oxygen consumption while walking at moderate speeds are very much less than would be predicted by extrapolation from running. There is no reason to consider that the energy cost of human walking is in any way remarkable.

Those remarks refer to locomotion on treadmills and similar firm, smooth substrates. White & Yousef (1978) measured the rates of oxygen consumption of men and reindeer (caribou) walking on roads and on tundra. On hard, well-packed roads, the two species had approximately equal net costs of transport. On soft but dry tundra the reindeer's costs increased by 13% and the men's by 60%. Unfortunately, no data for other species are available, so it is not known whether the reindeer is unusual in having its cost of transport so little affected by the change in terrain, or whether humans are unusual in being affected so much.

A few mammals make annual migrations over distances of several hundred kilometres (Baker, 1978). Wildebeest in the Serengeti, Tanzania, travel an annual circuit of about 1000 km. Caribou make annual migrations of up to 1000 km each way, between the forests of Alberta and Manitoba, and the tundra further north. These journeys may be compared with human hiking. The caribou travel 25–50 km per day (Baker, 1978). Similarly, some hikers travel the Pennine Way in 10 days, averaging 43 km per day, but guidebooks such as Hopkins (1989) recommend 19 days

(23 km per day) for average walkers. A Victorian walker (White, 1861) spent a month walking round Yorkshire, travelling 35 and 42 km on his two most strenuous days but 23–29 km on each of the others.

Home range

The home range of an animal is the area in which it generally moves around in the course of feeding, resting, reproduction, etc. It may travel further afield occasionally but spends most of the time in the home range. It is very difficult to define home range unambiguously, but estimates have been made for many species of mammal, as Figure 2.5 shows. Large mammals generally occupy larger ranges than small ones: for example, 9000 hectares (almost equivalent to a square of side 10 km) for 200 kg bears but only 60 hectares or 800 m square for 3 kg opossums. There are also marked differences between mammals of similar size, that seem to be related to diet. Carnivores generally occupy much larger ranges than herbivores, and omnivores have ranges of intermediate size.

In a discussion of physical activity, it would be pointless to estimate home ranges for humans who travel by car. For many centuries prior to the invention of cars, England had been divided into townships which may be regarded as the human equivalent of the home ranges of animals. A typical township consisted of a village together with its fields and the moor on which sheep and cattle were grazed. People would travel occasionally outside the township (for example, to market) but most worked in the township were they lived. Some townships included several villages or hamlets, each with its own fields, so were larger than the home ranges of most inhabitants, but the township nevertheless seems the best conveniently available estimate of human home range in Britain. Of townships in the West Riding of Yorkshire 317 had areas of 900 ± 811 hectares (mean and standard deviation). The mean value is equivalent to a square of side 3 km. Figure 2.5 shows that it is close to the predicted home range for an omnivorous mammal of 60 kg mass.

The Yorkshire population was an agricultural one. In contrast, the !Kung San (Bushmen) of Botswana are hunter–gatherers, and use much larger home ranges. The n!ore (the block of land around a water hole) is the basic subsistence area for the group living at the waterhole. Lee (1979) gives adult body masses for !Kung of 40–50 kg and estimates the area of n!ores as 30 000 to 60 000 hectares, about equal to the predicted home range area for a mammalian carnivore. However, in another publication Lee (1968) states that for all but a few weeks of the year, members of !Kung camps use areas of only 10 000 hectares. This is possibly the best estimate of their home range, and is shown in Figure 2.5.

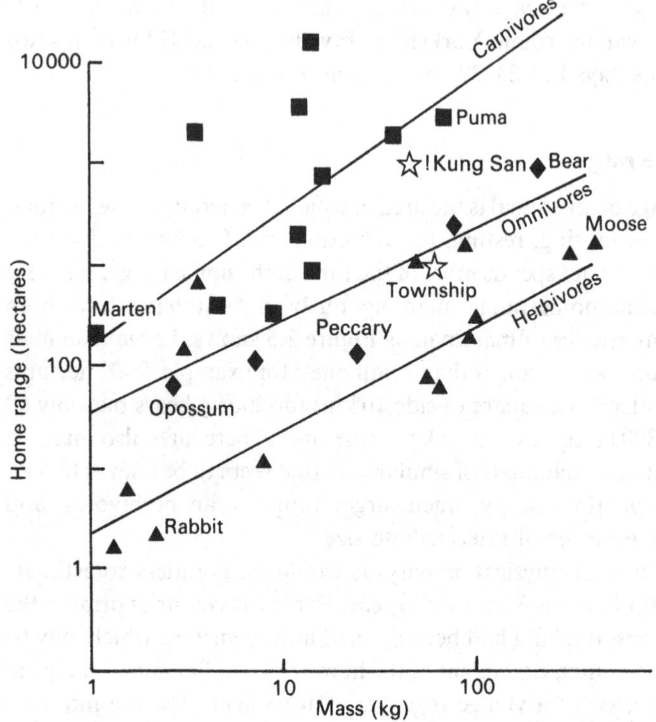

Figure 2.5. A graph on logarithmic coordinates of home range against body mass for carnivorous (squares), omnivorous (diamonds) and herbivorous mammals (triangles). All lines and points (except the point marked 'township') are from Harestad & Bunnell (1970), who compiled data for 53 species of North American mammals. 'Township' shows the mean of the areas of the 317 townships in the Wapentakes of Agbrigg, Barkston Ash and Claro in the West Riding of Yorkshire (Minchin, 1974). It is shown at a body mass of 60 kg to represent human home ranges prior to the introduction of motor vehicles. The point for !Kung San is explained in the text.

Conclusion

It has been seen that the basal metabolic rates of people lie in the mid-range for mammals of similar mass. Metabolic rates averaged over a week even for active people such as labourers are much lower than the field metabolic rates of deer and sea-lions, the only mammals of similar mass for which data are available. This does not necessarily mean that average metabolic rates are unusually low for mammals of human size; the basal metabolic rates of deer and seals are unusually high, so their field metabolic rates may also be high. Endurance athletes are capable of using oxygen much faster than most other people, but the range of variation among humans spans the rate predicted from animal studies, for mammals of our mass. No human athletes seem to be as far above the regression line of Figure 2.3 as are dogs and ponies, nor are more typical people as far below the lines as are cattle and sheep.

People cannot sprint as fast as greyhounds and racehorses, which have been bred specifically for speed. The limited and unreliable data that are available for wild animals make it seem probable that humans are slower sprinters than most mammals of similar mass. Humans perform better as endurance runners: maximum aerobic speeds of human athletes are much slower than those of dogs and horses but faster than those of cattle and sheep. People in cities walk very fast, but average walking speeds in villages are close to those of gazelles and wildebeest. Humans use more energy when we run than would be expected for mammals of our mass, but our walking is more economical. Hikers like to walk about as far in a day, as do migrating caribou.

The areas of townships seem reasonable estimates of the home ranges of British people, prior to the introduction of motor transport. They are smaller than the home ranges of most carnivorous mammals of similar mass, but larger than the ranges of most herbivores. !Kung San use larger home ranges.

Metabolic rates, speeds and home ranges may all be used as measures of human activity. Comparisons with animal data lead to a disappointingly dull conclusion. People are neither remarkably active for mammals of our size, nor remarkably inactive. Sprinting speeds are rather low but, in other respects, activity is much as might be expected.

There is an obvious explanation for humans being poor sprinters. Humans do not use arm or back muscles in running, except in a very minor way. For that reason, the mass of muscle that is available to power sprinting is probably much less than for most quadrupedal mammals of human size. This should not affect human performance as endurance runners, which depends more on the rate at which the respiratory and circulatory systems can deliver oxygen to the muscles, than on the mass of muscle.

In conclusion, human activity is unremarkable, but to be more specific: the points for human basal metabolism in Figure 2.1 are just a little above the line, as is that for a peccary (a wild pig). Figure 2.3 shows that the maximum aerobic metabolic rates of endurance for athletes and pigs are each a little above the prediction of the line. Warthogs can sprint at 10 m/s (Alexander *et al.*, 1977), about as fast as people. The maximum aerobic speed of domestic pigs, recorded by Weibel & Taylor (1981), is 4.6 m/s, a little lower than the human value of 6 m/s. It must be admitted that the home range for peccaries shown in Figure 2.5 is further below the omnivore line than is the human township area, but, by all the other measures of activity, people seem rather like pigs.

Acknowledgements

I am grateful to Dr Stanley Ulijaszek for information about the !Kung San.

References

Alexander, R.McN. (1984). Stride length and speed for adults, children and fossil hominids. *American Journal of Physical Anthropology*, **63**, 23–7.
Alexander, R.McN., Langman, V.A. & Jayes, A.S. (1977). Fast locomotion of some African ungulates. *Journal of Zoology* **183**, 291–300.
Alexander, R.McN. & Maloiy, G.M.O. (1989). Locomotion of African mammals. *Symposia of the Zoological Society of London*, **61**, 163–80.
Åstrand, P.-O. & Rodahl, K. (1986). *Textbook of Work Physiology*. 3rd Edn, New York: McGraw-Hill.
Baker, R.R. (1978). *The Evolutionary Ecology of Animal Migration*. London: Hodder & Stoughton.
Ballreich, R. & Kuhlow, A. (1986). *Biomechanik der Leichtathletik*. Stuttgart: Enke.
Blaxter, K.L. (1989). *Energy Metabolism in Animals and Man*. Cambridge: Cambridge University Press.
Bornstein, M.H. & Bornstein, H.G. (1976). The pace of life. *Nature*, **259**, 557–9.
Durnin, J.G.V.A. (1985). The energy cost of exercise. *Proceedings of the Nutrition Society*, **44**, 273–82.
Elliott, J.P., Cowan, I.McT. & Holling, C.S. (1977). Prey capture by the African lion. *Canadian Journal of Zoology*, **55**, 1811–28.
Garland, T. (1983). The relation between maximum running speed and body mass in terrestrial mammals. *Journal of Zoology*, **199**, 157–70.
Garland, T., Geiser, F. & Baudinette, R.V. (1988). Comparative locomotor performance of marsupial and placental mammals. *Journal of Zoology*, **215**, 505–22.
Harestad, A.S. & Bunnell, F.L. (1970). Home range and body weight – a reevaluation. *Ecology*, **60**, 389–402.
Hildebrand, M. (1959). Motions of the running cheetah and horse. *Journal of Mammalogy*, **40**, 481–95.
Hopkins, T. (1989). *Pennine Way North*. Ordnance Survey National Trail Guide 6. London: Aurum Press.
Huyssen, V. & Lacey, R.C. (1985). Basal metabolic rates in mammals: taxonomic differences in the allometry of basal metabolic rate and body mass. *Comparative Biochemistry and Physiology*, **81A**, 741–54.
Lee, R.B. (1968). What hunters do for a living, or how to make out on scarce resources. In *Man the Hunter*, Lee R.B. & De Vore I. (eds). pp. 30–48. New York: Aldine De Gruyter.
Lee, R.B. (1979). *The !Kung San*. Cambridge: Cambridge University Press.
Macdonald, D. (1984). *The Encyclopaedia of Mammals*. New York: Facts on File Publications.

Margaria, R. (1976). *Biomechanics and Energetics of Muscular Exercise*. Oxford: Clarendon Press.

Margaria, R., Cerretelli, P., Anghemo, P. & Sassi, G. (1963). Energy cost of running. *Journal of Applied Physiology*, **18**, 367–70.

McNab, B.K. (1986). The influence of food habits on the energetics of eutherian mammals. *Ecological Monographs*, **56**, 1–519.

Minchin, G.S. (1974). Table of population, 1801–1901. In *The Victoria History of the County of York*, Page W. (ed.), **3**, 485–548. London: University of London Institute of Historical Research.

Nagy, K.A. (1987). Field metabolic rate and food requirement scaling in mammals and birds. *Ecological Monographs*, **57**, 111–28.

Pennycuick, C.J. (1975). On the running of the gnu (*Connochaetes taurinus*) and other animals. *Journal of Experimental Biology*, **63**, 775–99.

Taylor, C.R., Heglund, N.C. & Maloiy, G.M.O. (1982). Energetics and mechanics of terrestrial locomotion. I. Metabolic energy consumption as a function of speed and body size in birds and mammals. *Journal of Experimental Biology*, **97**, 1–21.

Weibel, E.R. & Taylor, C.R. (eds) (1981). Design of the mammalian respiratory system. *Respiration Physiology*, **44**, 1–164.

White, R.G. & Yousef, M.K. (1978). Energy expenditure in reindeer walking on roads and on tundra. *Canadian Journal of Zoology*, **56**, 215–23.

White, W. (1861). *A Month in Yorkshire*. London: Chapman & Hall.

3 *Physical activity levels – past and present*

J.V.G.A. DURNIN

The problem about attempting a scientific discussion of 'physical activity levels – past and present' is that factual information is limited. Data which are valid generally, and which will tell us how active people were 20, 50 or 100 years ago, what sort of activity they indulged in, how it was influenced by socio-economic class or by age, and how this has altered at the present day, do not exist.

An attempt can be made to analyse the changing nature of occupations as far as physical activity is concerned, and the reasons why people might be active can be dissected out. Is activity enforced by the nature of the occupation or is it a freely chosen component of our leisure time? Or, indeed, do both sometimes operate together?

Although the reasons why some people are active are not a basic consideration of this paper, they obviously affect the level of activity and need to be discussed, even somewhat superficially. A basic assumption might be held that physical activity is natural and is beneficial for physical and psychological health at all ages.

Some reasonably persuasive evidence can be produced that such a view more or less fits in with the facts. The fact can be cited that many people obviously seem to enjoy being active, and the apparent benefits of activity can be enumerated with convincing facility: an increased feeling of well-being, improved cardio-vascular and muscular function, hopefully some protection against cardio-vascular diseases, maintenance of a certain degree of agility and joint flexibility, a greater likelihood of avoiding obesity and of developing a trim, muscular, and attractive physique, increasing energy expenditure with the advantage (particularly important for elderly people) that appetite and food intake will also be improved and therefore there will be less chance of nutritional deficiencies occurring.

Almost all of these statements are self-evident: they hardly need to be proved since no sensible person would dispute them. However, if we examine them one by one with a cynical eye, the evidence is less convincing. For example, there is no doubt that many, probably most, people we know who are active do seem to enjoy a *heightened feeling of well-being*, often proportionate to the intensity and induced exhaustion of the exercise. Whether this is a natural resultant of the activity, or whether a certain type of personality has to exist beforehand, perhaps related to an ambition to do better than others, or as a compensation for feelings of inferiority, is not known. The feeling of well-being after *exercise* (but perhaps not necessarily always after *physical activity*) seems to be shared by many people, but they might still represent a relatively small proportion of a normal population. Most people probably get this sensation most acutely after fairly severe exercise, although 'fairly severe' is a level with a great deal of individual variation. However, it is possible that a large majority of people do not derive this same satisfaction from activity, and in this category, many children, middle-aged, and perhaps most elderly men and women might be included.

There is a considerable problem in this context related to whether or not the activity is obligatory or chosen voluntarily. For example, it is doubtful if men who pedal bicycle rickshaws in the humid heat of Malaysia or Thailand, or poor Indian coolie labourers, or indeed the coal-face miners we studied in Scotland (Garry *et al.* (1955), gained any appreciable feeling of well-being from the physical activity they were obliged to undertake because of the requirements of their work.

This would seem to imply that it is not the physical activity itself which produces this particular beneficial effect, but the motivation of the individual. Children are an obvious illustration of the difficulty in making generalizations about 'well-being' resulting from activity; if the 'activity' consists of tasks around the home, garden, or farm, or long walks in the country, they may not associate activity with pleasure.

The second influence which might encourage people to be active was concerned with *improving cardio-vascular and muscular function*, but whether or not this conveys some *protection against degenerative cardio-vascular disease* is still a subject of some controversy. A few years ago, when the author was a member of the COMA Panel on 'Diet and Cardio-Vascular Disease' (DHSS, 1984), there was a series of meetings at which one or two speakers would review a pertinent topic and this would be followed by an exhaustive discussion. The author's subject was the influence of physical activity on cardio-vascular disease, including any interactions with diet. His conclusions, after reviewing the large, but largely very unsatisfactory epidemiological literature was that he was convinced that *some* protective action exists but not many members of the Panel were more than partially persuaded.

A further attractive benefit from activity is the *maintenance of a variable degree of agility and joint flexibility* and this is probably undisputed, but not entirely, and not in all circumstances. One is forced to admit that a very high risk of joint and tendon damage is associated with running on hard surfaces for lengthy periods over several years.

Another inducement to activity is that exercise has been suggested as *helping to avoid obesity and to developing a trim, muscular and attractive physique*. There is no doubt that at least parts of that philosophy have a considerable appeal to some people. People who are regularly active are less likely to be obese, and certain types of activity, such as weight-training, sometimes produce attractive, muscular bodies, although most non-participants in that 'sport' would feel that there is a limit beyond which muscular development results in a grotesque distortion of the human body.

However, if the relationship between activity and obesity is analysed, there are remarkably few papers in the literature which convincingly demonstrate the efficacy of *treating* obesity by exercise. Most of the evidence is either anecdotal or subjective. Everyone has experience of seeing an acquaintance (or hearing about someone) who, through regular exercise such as jogging, has changed his physique from that of a fat slob to one of acceptable shape, and simultaneously, sometimes, has become a sexually aggressive individual. Such impressions are not very helpful in scientifically assessing the situation in the community.

Possibly one of the most important roles of physical activity is in *increasing energy expenditure in the elderly*. Malnutrition, in the developed industrialized world, excluding the severely impoverished, the poorly looked-after patients in institutions, and peculiar food faddists, is virtually non-existent except for some groups of elderly people. Elderly people living alone or in depressed circumstances, or lacking in appetite and uninterested in food, can sometimes show signs or symptoms of malnutrition. If they can be encouraged to be more active, particularly by walking more, joining social clubs, bowling clubs, etc, the increased energy expenditure will inevitably lead to a fuller appetite, to a larger food intake, and thus a diminished chance of poor nutrition. Whether or not these attributes of activity, now fairly generally accepted in the medical field, have resulted in increased levels of activity, is problematical and unknown for the general elderly population.

Definition of 'levels'

Brief reference has been made to the obligatory or voluntary aspects of activity, and a distinction has been made between the physical and psychological rewards. It is perhaps arguable that it may not always be the case that no psychological benefit accrues from physical activity which has been forced upon the individual because of the nature of the occupation. Certain types of activity undertaken in pleasant surroundings, such as non-mechanized farming or some forms of forestry work, certainly may induce a feeling of well-being, as well as conferring the other physical benefits.

However, independently of the possible distinctions between occupational and leisure activities, 'levels of physical activity' may be considered in different ways. 'Levels' could refer to the popularity of activity, i.e. how common it was or is within different populations, and these populations might be British or American or Indian or Eskimo. It is not possible in a short chapter to consider such a global approach so reference here will be restricted to UK populations. Nevertheless, even with this limitation, there may be marked differences between urban and rural groups, there may be an influence due to socio-economic class, age may be an important determinant, and there may be variations due to gender (although feminists may dispute this last suggestion). Reliable valid information on the influence of these variables is virtually unknown.

Apart from the general population aspect of activity, the type of activity, e.g. walking or jogging or playing sports, its strenuousness, and its frequency may all differ and, in turn, these also may be affected by the variables listed above.

'Levels' of activity may also refer to the actual energy cost of being active, particularly with reference to the demands of occupational activity. There is some evidence to suggest that the various categories of energy expenditures required by different types of work have not really altered in the past 50–100 years. For example, Voit, in 1881, estimated the energy value of the food consumed by labourers in Germany, as having an average value of 3055 kcal/d. Langworthy, in a review of *Food and Diet in the United States* published in 1907, states that, after reviewing all the diet studies in the USA and in other countries, 'the results obtained the world over do not differ markedly from a general average of 3000 calories of energy'. And, in a publication from the US Dept. of Agriculture in 1939 on Energy Requirements, five different categories are listed (Morey, 1939).

- Sedentary 2400 kcal/d
- Light work 2700 ,,
- Moderate work 3000 ,,
- Hard work 3600 ,,
- Very hard work 4500 ,,

Apart from 'hard work', which is very uncommon in western society nowadays, and that of 'very hard work', which must have been non-existent for decades, these values are still appropriate for the present day. Therefore, the energy levels of different categories of physical activity have probably not altered much; the problem is that little is known, in quantitative terms, about how the *prevalence* of these levels has changed in the population over the years. It can be assumed with reasonable confidence that the proportion of people who need to be physically active because of the nature of their occupation has been steadily diminishing for the past 30 or 40 years and, at the present time, apart from workers in the construction industry, and men and women in some branches of the Armed Services, occupation does not exert much of an effect on increasing physical activity. There are census data on the numbers of people employed in various occupations showing how relatively few people have energetic occupations, and an indirect influence of this lack of enforced activity can be detected by the steadily increasing prevalance of obesity. It is also quite illustrative of this trend to examine old documentary films, taken 50 or 60 years ago, showing people walking in the streets or men coming out of the factory gate: if one looks carefully at these scenes an obese person is rarely seen. Equivalent film sequences at the present day hardly give a similar impression. In the most highly developed industrialized country, the USA, any random documentary will almost inevitably demonstrate frequent and gross obesity. An almost extreme example of this could be witnessed in a series of short films on BBC television concerned with the poor socio-economic conditions of people living in the Appalachians, including statements suggesting that some of these people were so poor that they could not purchase enough food; and many of the interviewees were so grossly fat as to be almost pathologically obese!

Two factors are markedly connected with changing levels of physical activity, both of them operating particularly in children. These are television and transport. Everyone nowadays spends hours per day watching television, and few people walk as much as in the past, because of the availability of transport by car, bus, and train. These are dogmatic statements with little objectivity to support them, but it is at least interesting to note the information recorded in the Nielsen 'Report on Television' in the USA in 1990 that 2–4 year-old children watch TV on average 4 h daily and 6–11 year-olds 3.5 h daily (Nielsen Media Research, 1990). This considerable amount of physically inactive time might naively be justified as having perhaps some educational benefits, but it undoubtedly has some serious disadvantages physically.

Those two influences almost certainly contribute largely to the explanation for the progressive decrease in energy intake of adolescents in the UK from the 1930s up to the present day (Table 3.1): these intake data are of great interest since, by chance, the body weights of both the boys and the girls happened to be almost identical within each sex for all the four groups. There is thus no confusing influence of differing body weights. The methodology was the same on each occasion. The only conceivable

Table 3.1. *Energy intakes of 14–15 year-old boys and girls in the United Kingdom (kcal/d)*

	Boys	Girls
1930s (ref. 1)	3065	2640
1960s (ref. 2)	2795	2270
1970s (refs 2, 3, 4)	2610	2020
1980s (ref. 5)	2490	1880

References:
1. Widdowson (1947).
2. Durnin *et al.* (1974).
3. Cook *et al.* (1973).
4. DHSS, unpublished observations.
5. D of H (1989).

explanation for the very marked reduction in energy intake, which must reflect diminished energy expenditure since the numbers of children studied was quite large, is that physical activity has also radically decreased. Some justification for this belief comes from the findings of Armstrong *et al.* (1990) on the low levels of habitual physical activity of 11–16 year-old children in Devon.

Voluntary, as opposed to obligatory, activity has become the major determinant of the overall activity pattern for the great majority of people in our type of society. As a cynic, one might suppose that, because a choice is now required for activity, this would inevitably result in a progressively more inactive way of life. However, cultural and, possibly, health influences affect the pattern. The 'health' motivation is obviously a major one with so much emphasis coming from the media on healthy life styles. As an illustration of a cultural influence, boys attending certain English public (in reality, very private) schools might be exposed to considerable pressure and encouragement to be physically active on a fairly frequent and regular basis.

Another cultural factor, perhaps contributing a powerful influence on women, might be the social pressure to be slim and physically attractive. Perhaps this, more than any other single factor such as 'health', has been responsible for the apparently remarkable increase in aerobics and in jogging in women. But jogging is of endemic proportions for both sexes. Surely, at no time in human history, has such useless and selfish exercise been undertaken by so many people. Nothing constructive is accomplished, no general improvement in anything has occurred, except of a temporary nature, and of concern only for the individual. When jogging is considered as the form of physical activity which has been chosen by large numbers of individuals, many of them otherwise sensible and perhaps even intelligent, caring people, it is depressing to think of this as a sign of the degradation to which our industrialized culture has descended. Surely something more constructive could be chosen: gardening, playing some physically or psychologically competitive game, climbing hills, walking through a park or in the country, bicycling, even playing golf, anything other than this peculiar form of exercise.

However, perhaps this lamentable form of physical activity is not as prevalent as it seems. In real life, it is extremely difficult to obtain reliable data about habitual levels of physical activity in any large-sized population. Typical of this difficulty of interpretation in the participation in activity is the interesting information given by Collins (1987) in his paper on 'the Sociology of sport and exercise in the UK', which is entirely qualitative and impossible to use in any precise way related to frequency, degree, or exact type of activity. The physiologist or epidemiologist gains very little from such sociological data.

The main problem is that there is often no such thing as any consistent level of physical activity. Many people, at some periods of their life, may suddenly become concerned that their lifestyle is lacking exceedingly in physical activity. They may identify activity in this context as pertaining only to rather strenuous physical exercise and, perceiving that they rarely do anything really energetic, take up some highly artificial form of exercise, indulge in it to extreme levels for very limited periods of time, and regress to their previous level of lassitude within a matter of months or a year or two.

There are innumerable examples in the nutritional literature of energy intakes in the diet which are at such a low level that only minimal physical activity would be possible (Durnin, 1990). It can be concluded legitimately that levels of physical activity for the general population are lower now than they were in the past, mainly because there is seldom any longer an absolute requirement for men to be physically active in order to stay alive. This means that levels of physical activity now are dependent on voluntary decisions and since there seems to be no innate drive for man, in his natural state, to indulge in activity of other than modest degree, levels of physical activity are likely to remain low for most people. In a socio-economic environment where food is plentiful and where, physiologically, people have not been imbued with any particularly effective control of their appetite, there are obvious dangers attached to this lifestyle.

References

Armstrong, N., Balding, J., Gentle, P. & Kirby, B. (1990). Patterns of physical activity among 11 to 16 year old British children. *British Medical Journal*, **301**, 203–5.

Collins, M. (1987). Sociology of sport and exercise in the United Kingdom. In *Exercise, Heart, Health*, pp. 89–102. London: Coronary Prevention Group.

Cook, J., Altman, D.G., Moore, D.M.C., Topp, S.G. & Holland, W.W. (1973). A survey of the nutritional status of schoolchildren. *British Journal of Preventive and Social Medicine*, **27**, 91–9.

Department of Health and Social Security (1984). *Diet and Cardiovascular Disease*. Report on Health and Social Subjects 28, London: HMSO.

Department of Health (1989). *The Diets of British Schoolchildren*. Report on Health and Social Subjects 36, London: HMSO.

Durnin, J.V.G.A. (1990). Is satisfactory energy balance possible on 'low' energy intakes? *Proceedings of the Nutrition Society*, India, **36**, 1–9.

Durnin, J.V.G.A., Lonergan, M.E., Good, J. & Ewan, A. (1974). A cross-sectional nutritional and anthropometric study, with an interval of 7 years, on 611 young adolescent schoolchildren. *British Journal of Nutrition*, **32**, 169–79.

Garry, R.C., Passmore, R., Warnock, G.M. & Durnin, J.V.G.A. (1955). *Studies on Expenditure of Energy and Consumption of Food by Miners and Clerks, Fife, Scotland, 1952*. London: MRC Special Report Series No. 289.

Langworthy, C.F. (1907). Food and diet in the United State. *US Department of Agriculture Year Book*, pp. 361–78. Washington.

Morey, N.B. (1939). Energy requirements. US Department of Agriculture *Food and Life*, pp. 160–72. Washington: Superintendent of Documents.

Nielsen Media Research (1990). Nielsen Report on Television. New York: 1290 Avenue of the Americas.

Voit, C. (1881). Physiologie des Stoffwechsels. In *Handbuch der Physiologie*. Leipzig, H.L. ed., p. 519.

Widdowson, E.M. (1947). A study of individual children's diets. *Medical Research Council Special Report Series* No. 257. London: HMSO.

4 *The validity of health measurements*

J.R. KEMM

What is health?

Health is a term that is widely used (for example, in the title of this symposium) but the underlying assumption that readers share a common understanding of its meaning is rarely justified (Seedhouse, 1986). Before the validity of health measurements can be discussed, the meaning of health must be clarified. A useful starting point is the definition of health included in the charter of the World Health Organization (WHO, 1946): 'Health is a state of complete physical, mental and social well being and not merely the absence of disease or physical infirmity'.

As a statement of ideals it cannot be faulted, but it is doubtful that it describes an attainable goal and it does not offer a useful basis for measurement. A different definition is 'Health is a state in which a person is enabled to work to fulfil their realistic chosen and biological potential' (Seedhouse, 1986). From a practical point, this is much to be preferred, recognizing as it does that different individuals have different potentials. However, for purposes of measurement, it complicates things even further requiring not only the present state of each individual to be assessed but also their theoretical potential.

None the less, three important characteristics of health emerge from these definitions:

Disease is absent.

Other features ('positive health') are present.

The concept is multidimensional.

Broad definitions of health would justify the inclusion of indices such as crime rates, literacy rates, religious observance and voting behaviour in measures of the health of populations. However, this chapter will consider only those measures related to physical function, mental function or feelings.

The uses of health measurement

Health measures may be applied to individuals or to populations. Measures of population health can be derived from measures of individual health by aggregation, but the converse is not true.

Table 4.1. *Measures of health*

Measures of health
Absence of disease
Death rates and Life expectancy
Morbidity rates based on service utilization
Disease frequency rates from special surveys
Measures of risk status
Health of lifestyle assessment
Absence of ill health
Prevalence of symptoms and self-rated ill-health
Measures of psychiatric distress
Presence of 'positive health'
Measures of fitness
Other measures of positive health

Measures of individual health may be used to identify those with health problems which merit further investigation, to assist diagnosis, to inform patient management decisions and to monitor progress. Measures of population health are used to assess need, to determine priorities, to evaluate the effect of health care programmes, and to study the correlates and possible determinants of health. This chapter will be concerned primarily with this last purpose.

Measures of health can be categorized as shown in Table 4.1. Most of them are measures of ill-health rather than health but their inclusion is justified on the grounds that absence of ill-health is a necessary, but not sufficient, feature of health.

Mortality rates

Death may be equated with the total absence of health, and one characteristic of healthy populations is delay of death to an old age. In comparing death rates between populations, due allowance has to be made for age structure since other things being equal crude death rates will be higher in populations which contain higher proportions of older people. There are various ways of making this allowance. With large populations, the most informative approach is to examine age-specific death rates though this has the disadvantage that the population cannot be described with a single number. Age standardization allows one to discount differences in age structure between populations, but age standardized rates will chiefly reflect death rates in the oldest age-range since this is when most of the deaths take place. Standardized Mortality Ratios (SMRs) are one commonly used form of age standardization. The fact that different

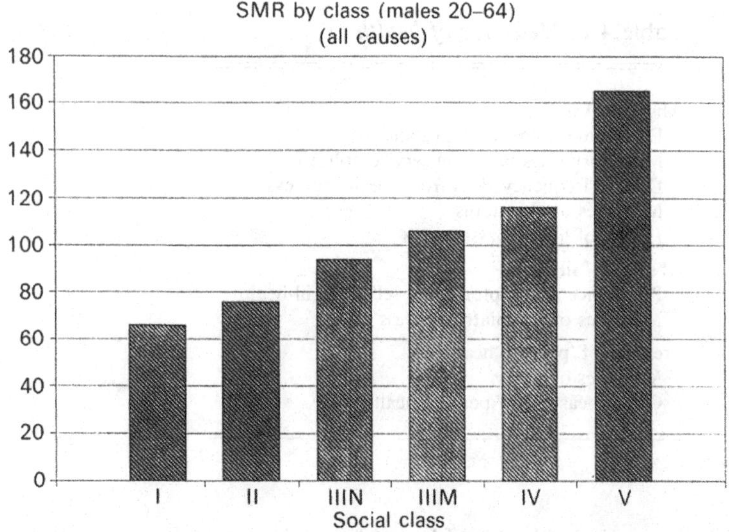

Figure 4.1. Social class gradients in standardized mortality ratio males (England and Wales) 1979–82.

social classes have different SMRs (Figure 4.1) suggests that they also enjoy different levels of health.

It may be argued that deaths at a young age (say 20) are more indicative of an unhealthy population than deaths at an older age (say 80). Standardized death rates make no allowance for this but potential years of life lost (PYLL) is a measure which takes it into account and attaches greater weight to early deaths than later ones. The simplest way to calculate PYLL is to assume that all have a life expectancy of 70 so that a death at age 69 represents one PYLL while one at age 20 represents 50 PYLL. There are more refined ways of calculating PYLL (Romeder & McWhinnie, 1977).

Another way of attaching more weight to early deaths is calculation of life expectancy. The name of this value suggests that it is a true indicator of 'positive health' but it is merely a variant of death rate, being calculated from age-specific death rates. Figure 4.2 shows how life expectancy in this country has changed over the years, and invites the inference that health has improved.

While death rates reflect length of life, they tell nothing about its quality. It is easy to imagine two populations with identical death rates but differing widely in the prevalence of non-fatal disease.

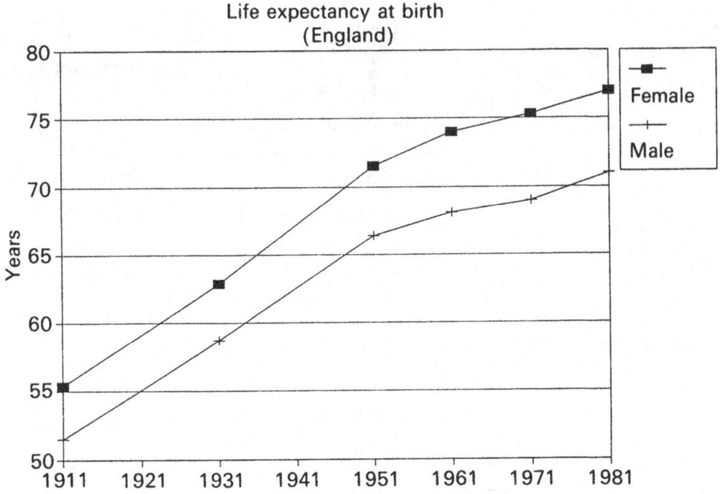

Figure 4.2. Changes in life expectancy in England 1911–81.

Disease frequency – service data

Frequency of disease may be used as another inverse measure of health, it being assumed that a low frequency of disease is indicative of a healthy population. As with death data, difference in age structure of the population has to be allowed for either by quoting age-specific rates or age standardized rates.

Many estimates of disease prevalence are based on service data such as hospital discharge rates or primary care consultation rates. It is important to realize that these are influenced by many factors other than the frequency of the conditions in question.

The process through which a health event is registered in service statistics involves many steps at each of which errors or biases can enter (Figure 4.3). Service variables such as diagnostic fashion, admission and treatment policies, availability of resources and competing demands for those resources will all have major effects on apparent disease frequency. Failure to record, coding errors and data entry errors may all add to the inaccuracy of such measures. Patient behaviour will also influence apparent rates based on service data. Even with serious conditions, many patients do not present for treatment and a change in apparent frequency of the condition may merely reflect increased readiness to present.

A study of the prevalence of alcohol-related illness illustrates how routine service data can mislead. Admission rates to psychiatric hospitals

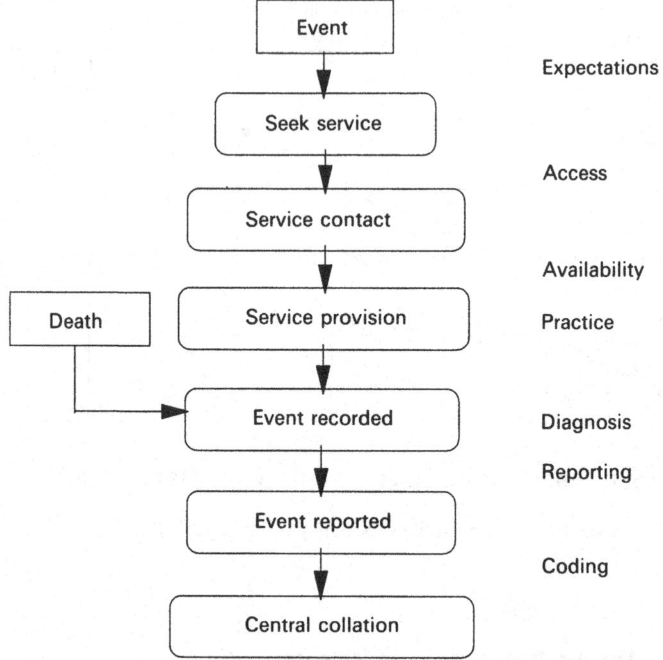

Figure 4.3. Steps in registering of a health event.

with this diagnosis have been noted to be higher in Scotland than in England, with the implication that this condition was more prevalent in Scots. A comparison of admission rates for two Scottish districts (Highlands and Tayside) with those for an English district (South East Kent) confirmed this difference. However, more detailed examination using information not available from routine sources showed that the frequency of this diagnosis was not very different in the three districts, and that the difference in psychiatric admission rates was largely attributable to greater provision of psychiatric beds and different treatment policies (Latcham & Kreitman, 1984).

A further problem with hospital data is that it is event rather than person based so it is not possible to distinguish between one person admitted 50 times in a year or 50 people admitted once.

If hospital data is unsatisfactory the data from general practice (primary care) is even more problematic. The Office of Population Census and Surveys has run several National Morbidity Surveys (RCGP, OPCS & DHSS, 1986). There are also various national and regional schemes which collect and report consultation data from a network of sentinel practices for communicable and other diseases (Flemming *et al.*, 1988). All these schemes record from selected volunteer practices and, although the coverage seems adequate, this must raise questions about the representativeness of the data. The coding systems used make allowance for far more conditions than does the International Classification of Disease (ICD) but standardization of diagnostic habit between participating practices must be difficult.

Disease prevalence – special surveys

The problems of routinely collected service data mean that, for most conditions, the only way in which reliable prevalence and incidence data can be obtained is by special population surveys. Such surveys require the use of properly standardized diagnostic criteria and then obsessive attention to methods of ascertainment.

Risk factor status

All measures discussed so far have been measures of obvious ill-health. Another approach is to assess the risk of developing disease. This has been most extensively developed for ischaemic heart disease but could be applied to other conditions. Risk factors for ischaemic heart disease (IHD) include high serum total cholesterol, high serum low density lipoprotein (LDL) cholesterol, low serum HDL cholesterol, raised blood pressure, smoking, impaired glucose tolerance and obesity. These, and other risk factors, can be measured and their distribution in the population described. Schemes have been devised to combine several different risk factors into risk scores which are global measures of risk of developing IHD (Shaper *et al.*, 1986). The validity of the risk factor approach hinges on the validity of the underlying assumption that these factors increase the probability of developing the condition under consideration. Measures of risk cannot simply be transferred between cultures. There is, for example, a suggestion that risk scores developed for Anglo-Saxon populations may not be appropriate for use in groups originating from the Indian subcontinent, since conventional risk factors fail to account for the high SMRs observed in that group (Coronary Prevention Group, 1986).

Measures of healthy lifestyle

A healthy population is one which engages in a healthy lifestyle. This statement is tautologous but it suggests another approach to the measurement of health. Certain lifestyles such as non-smoking, eating particular nutrient patterns (low fat, low saturates, low salt, high fibre, high complex carbohydrate), eating particular food patterns (whole cereal foods, fruit and vegetables, fish, etc), moderate consumption of alcohol, safe sexual practices, regular exercise, adequate relaxation periods and so on might be defined as 'healthy'. The prevalence of these lifestyles in the population can then be measured.

Most lifestyles currently described as healthy have been so characterized on the grounds that they reduce risk of subsequent disease. In theory, it should be possible to identify 'healthy' lifestyles which promote 'positive' health, but for the time being lifestyle measures should be seen as related to the absence of disease rather than the presence of 'positive' health.

The validity of using lifestyle measures depends on the assumption that those who adopt them are healthy. This whole approach has been characterized as 'healthist' with the suggestion that there is something profoundly unhealthy about obsession with these aspects of lifestyle (Crawford, 1980). Even their association with low risk of disease has been questioned. Understanding has to be developed much further before lifestyle measures can safely be accepted as a proxy for measures of health.

Frequency of ill-health

All measures discussed so far are based on the notion of disease, a complex of symptoms, signs and pathological processes seen from the perspective of the health professional. This, however, is a very incomplete notion of ill-health, and excludes most of the aches, pains, impairments and disabilities which hinder people's enjoyment of life. The old joke about the patient who has been given 'a clean bill of health' by his doctor but feels awful is all too recognizable.

Numerous instruments have been designed to measure these aspects of ill-health and a selection of them are reviewed in the excellent monograph by McDowell and Newell (1987). Some instruments require rating by professionals while others rely on self-rating. Since it is the individual's perception of his/her own health which is important, it could be argued that self-rated instruments are to be preferred. However, self-rating must involve some loss of objectivity.

The effects of disease may be considered at three levels: impairment, disability and handicap. Impairment describes the limitation of function in anatomical or physiological terms, disability describes the restriction of ability to perform certain tasks, while handicap describes the interference caused in the person's role (Ibrahim, 1990). Most measures concentrate on the disability level. Questions on disability scales may focus on capacity (can the individual do this task?) or performance (does the individual do

Table 4.2. *Grogono–Woodgate Index for measuring health*

1. Work: normal, impaired or reduced, prevented
2. Hobbies and recreation: normal, impaired or reduced, prevented
3. Is patient free from malaise, pain or suffering?
4. Is patient free from worry or unhappiness?
5. Does patient communicate satisfactorily?
6. Does patient sleep satisfactorily?
7. Is patient independent of others for acts of daily living? (e.g. washing, feeding, dressing and moving)
8. Does patient eat and enjoy his food?
9. Is micturition and defaecation normal?
10. Has patient's state of health altered his sex life?

this task?). Several measures of disability have been developed for use with elderly patients or with patients suffering serious disabling disease.

The Arthritis Impact Scale (AIMS) and the Health Assessment Questionnaire (HAQ) developed for assessing function in patients with arthritis are examples of a self-rated disability scale (Spitz & Fries, 1987) while the Barthel scale (Mahoney & Barthel, 1965) for assessing elderly patients is an example of a professional rated scale. These scales were developed for use with ill patients and are generally unsuitable for use in population surveys since they fail to detect minor degrees of disability. A disability scale suitable for use with relatively minor degrees of disability was developed for use in the recent national survey of disability in Great Britain (Martin, Meltzer & Elliot, 1988).

Other measures concentrate on the interference with daily life, and cover symptoms as well as disabilities. The Grogono–Woodgate scale covers ten areas (Table 4.2) and is an early example of a professional rated scale designed for use with relatively dependent groups (Grogono & Woodgate, 1971). The Sickness Impact Profile (SIP) developed in the USA (Bergner *et al.*, 1981) but validated for use in the UK (Patrick *et al.*, 1985) is a self-rated instrument comprising 136 items covering 12 aspects of life. The Nottingham Health Profile is a much shorter instrument developed in the UK. (Hunt, McEwan & McKenna, 1986). None of these instruments detects very minor degrees of ill-health so that, when applied to general populations, most individuals will be rated as complaint free.

Mental distress

Feelings such as anxiety, sadness and fear are just as much ill-health as physical symptoms, but are not adequately covered by some of the physical symptom measures. A range of scales has been developed to quantitate the extent of psychiatric symptoms, some for specific conditions and others for a wide range of symptoms. One example of such a scale is the General Health Questionnaire (GHQ) (Goldberg & Hillier, 1979). In its initial form, it consists of 60 questions, but abbreviated versions with 30, 28, 20 and 12 questions have also been derived. Various subscales within the GHQ were identified including somatic symptoms, anxiety and insomnia, social dysfunction and depression. Other instruments explore quantity and quality of social interaction.

Measures of 'positive health'

Three classes of variable can be identified with regard to health. First, there are dichotomous variables such as dead/alive, smoker/non-smoker where one state is 'healthy' and the other 'not healthy'. Secondly, there are unipolar variables such as degree of pain where a zero magnitude is neutral (neither healthy nor not healthy) and increasing magnitude is increasingly not healthy (or healthy). Finally, there are bipolar qualities such as susceptible/resistant or weak/strong where increasing value represents a transition from not healthy to healthy. Most of the measures discussed so far can be categorized as unipolar measures of disease or ill-health rather than measures of 'positive' health.

There is no philosophical difficulty in accepting that states of health include something which is more than the absence of negative features. Happiness is more than the absence of sadness, and ease is more than the absence of distress or disease. The practical problem lies in defining that something, and then devising methods of measuring it.

Some authors (for example, Catford, 1983) have classed indicators of lower risk as 'positive' health indicators but since risk is conceptually linked to disease this does not seem satisfactory. Others have claimed the obverse of dichotomous indicators as 'positive' indicators. However, quoting not-smoking rates rather than smoking rates hardly advances understanding.

Physical fitness is one of the few measures which can confidently be classified as positive health, but will not be discussed further in this chapter since it is covered elsewhere in this volume.

Some genuine measures of 'positive' health are to be found in the subscales of the mental health questionnaires. The affect balance scale (McDowell & Praught, 1982) claims to measure 'positive affect' which is a feeling of optimism and contentment (Table 4.3) and the General Well-Being schedule (described in McDowell & Newell, 1987) has subscales called 'positive well-being', 'self-control' and 'vitality'. Measures of qualities such as self-esteem (Rosenberg, 1965) could also be included as

Table 4.3. *Affect balance scale (positive affect items)*

During the past few weeks did you ever feel:
particularly excited or interested in something?
proud because someone complimented you on something you had done?
pleased about having accomplished something?
on top of the world?
that things were going your way?

measure of 'positive' health. These elements are part of 'positive' health, but much more is needed to provide a fully satisfactory measure of the construct.

Global health measures

The problem of finding measures of health has been tackled by breaking the concept down into its constituent dimensions. Many of the instruments mentioned in the preceding sections such as the sickness impact profile, the Nottingham health profile and the general health questionnaire provide a range of subscales which are more informative than the overall scale.

However, subdivision into ever smaller and better defined dimensions takes us away from the overall construct of health. For some purposes, it is desirable to produce a global measure of health. This has been particularly stimulated by the economists' interest in measuring health 'utility'. Grogono and Woodgate (1971) suggested that their scale could be used in this way by simply summing the scores on the different items. Now, much more refined ways of producing global measures have been described.

One such measure is the quality adjusted life year (QALY). Different health states can be defined in terms of level of disability and discomfort and each state is then valued on a common scale, so that a year lived in perfect health has a QALY value of one, death has a QALY value of zero and years lived in varying degrees of disability or discomfort have QALY values between 1 and 0. (Some states such as being bed-bound in great discomfort are deemed to be less desirable than death and thus have a negative QALY value.) The quality of well-being scale (described in McDowell & Newell, 1987) involves a similar approach.

Construction of health measures

Indices based on death or prevalence of disease are supposed to be founded on firm diagnostic criteria and objective observation, often supported by chemico-physical measurements such as ECGs, X-rays and blood chemistry. The reality is often less secure but, in theory, the validity of these

Table 4.4. *Nottingham health profile sleep items and their weightings*

I take tablets to help me get to sleep	22.4
I'm waking up in the early hours of the morning	12.6
I lie awake for most of the night	27.0
It takes me a long time to get to sleep	16.1
I sleep badly at nights	21.7

measures does not seem to be a problem and will not be discussed in this chapter.

On the other hand, measures derived from responses to questions and subjective assessment seem, at first sight, to be much less trustworthy. The rationale for their construction and their performance has to be carefully scrutinized. There are three main stages in the construction of a measure describing the different dimensions, scaling each dimension and combining different dimensions to produce a summary measure.

Identifying the different dimensions

Scales are often constructed from several items. Underlying theory provides the rationale for grouping of a set of items, but procedures such as Guttman scaling, and Factor analysis should be used to test that the component items of a scale really are located on a single dimension. The items from the sleep subscale of the Nottingham health profile (Table 4.4) illustrate how a scale may be made up from several items. It should be noted, however, that there is some evidence that one of the items in this example (the tablet item) lies on a different dimension to the others (Kind, 1982).

Scaling ordinal responses

Many items use responses which form an ordinal scale. If the responses to different items are to be combined, this ordinal scale has to be converted into a scale which has at least interval properties. In many instruments, the values assigned to scale points are arbitrary. The Grogono–Woodgate index for example has responses normal impaired and incapacitated which are simply assigned the values 1, 0.5 and 0. The general health questionnaire has responses such as better than usual, same as usual, worse than usual and much worse than usual which were assigned the values of 1, 2, 3 and 4. Neither of these examples is likely to produce a truly interval scale.

The problem of converting ordinal to interval scales may be avoided by using dichotomous items. The general health questionnaire responses are sometimes effectively collapsed into two by assigning values of 0, 0, 1 and 1 to the four responses (Goldberg & Hillier, 1979).

Forming multi-item scales

Many health scales are made by combining the responses to several dichotomous items. The simplest approach is to give all items equal weight and sum them. Examples of this are to be found in the affect balance scale, the general health questionnaire and the health assessment questionnaire.

Other instruments, for example, the Nottingham health profile give different weight to each item (Table 4.4). In this example, the weights were derived by paired comparisons but panel ratings (used in the sickness impact profile) and other methods may also be used (Kind, 1982).

Several different ways of constructing multi-item scales can be found among the subscales of the survey of disability in Great Britain (Martin *et al.*, 1988). Most of the subscales are scored by taking the value of the most severe positive item. One subscale (consciousness) is scored by adding the weights of each item present while another subscale (intellectual) is scored by counting the number of positive items.

Non-dichotomous items (more than two scale points) provided they constitute interval scales can also be combined into multi-item scales. Martin (Martin *et al.*, 1988) used a model derived from panel ratings to form an overall disability scale combining the eleven disability subscales. The values ascribed to different combinations of distress and disability in the QALY scale were derived from panellists' ratings (Rosser & Kind, 1978). Other methods of comparing different health states are the time trade-off method and the standard gamble (Capewell, 1988). The time trade-off method requires panellists to estimate what duration of one outcome (e.g. perfect health) was of equal value to a year of another outcome (e.g. in pain and confined to bed). The standard gamble method uses a similar trade-off in which the probability of different outcomes is varied.

It is debatable to what extent combining measures of different dimensions into a global health measure is meaningful or useful. Several instruments such as the Nottingham health profile and the sickness impact profile are described as profiles and usually presented as a set of subscores (see Figure 4.4) rather than as a single score. When the different subscores are combined, a great deal of information is lost.

The difficulty of ascribing weights to the different dimensions increases as they become dissimilar. For example, it is easier to compare the relative

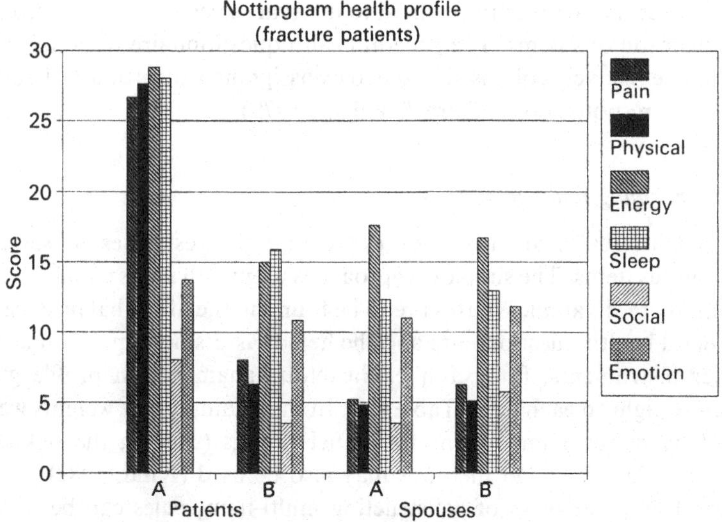

Figure 4.4. Construct validation of Nottingham health profile scores in fracture patients and their spouses: (A) soon after fracture; (B) after fracture was healed.

severity of inability to see and inability to hear than it is to compare the relative severity of physical disability and mental distress.

Validation of health measures

Content validity is the extent to which the measure covers all components of the construct being measured. Most health measures are deliberately limited to a small part of the construct of health. They therefore have low content validity as global measures of health though they may be content valid for a subset of that construct.

Face validity is the extent to which it appears reasonable that the instrument would measure the construct. Consideration of these two aspects of validity has been implicit in the discussion so far and will not be discussed further.

Instruments with low reliability cannot have good validity. Reliability of many of the instruments discussed has been measured by test–retest and found to be adequate.

Criterion validation

Criterion validation involves comparing the results of a method against those of another method (a gold standard) which is believed to produce

Table 4.5. *Criterion validation of the sickness impact profile*

Correlation of SIP scores with different clinical measures of disease severity in patients with rheumatoid arthritis

	Sickness impact profile scores		
Indicators of disease severity	Overall score	Physical sub-score	Psycho-social sub-score
Duration of RA	0.26*	0.36**	0.03
Morning stiffness	0.23*	0.15	0.32**
Erythrocyte sedimentation rate (ESR)	0.36**	0.44**	0.11
Anatomic stage	0.17	0.31**	−0.01
Mental problems	0.11	−0.03	0.27**

*$P<0.05$, **$P<0.01$.

correct results. Such validation is usually not possible for health measures since no gold standard measure is available. Instead, a new instrument will be compared with an existing instrument intended to measure something similar but slightly different. In such comparisons, it is not clear what degree of correlation between methods is indicative of satisfactory performance.

An example of criterion validation was a comparison of the sickness impact profile with various clinical measures of severity of illness in patients with rheumatoid arthritis (RA) (Table 4.5) (Deyo *et al.*, 1983). The physical subscore correlated with duration of disease, anatomical grading of severity, and ESR (an indicator of disease activity). The psycho-social subscore correlated with a clinical rating of mental problems. Criterion validation of the general health questionnaire against clinical rating in a standardized psychiatric interview has similarly shown that instrument to have reasonable criterion validity (Goldberg & Hillier, 1979).

Construct validation

Construct validation involves testing the extent to which results obtained with an instrument conform with hypotheses about the properties of the construct being measured. For example, it would be expected that people who had just sustained fractures would be less healthy than their spouses and that their health would improve as their fractures healed. This was the basis of a construct validation of the Nottingham health profile (Figure 4.4) which indeed showed high scores in the fracture patients which fell when the fracture healed and lower scores in their spouses which did not change (McKenna *et al.*, 1984). Studies such as these provide reassurance that the purported measures of health described in this chapter are, indeed, measuring health.

Cross-cultural studies

Health measures need to be validated for each population in which they are used. It is not safe to assume that, because a measure is valid with one population, it will be valid in one living in a different part of the same country, let alone one living in other countries with different languages, different cultures and different views of health. These problems of cross-cultural validation have been discussed by Hunt (1986). None the less several of the instruments mentioned in this chapter have been validated for use in different cultures.

Value judgements

Scientists like to believe that their work is objective and value-free but, in the measurement of health, value systems cannot be avoided. Value judgements are implicit in the definition of health, the selection of dimensions to be included in its measurement, and the relative weightings attached to the different components. Even where these decisions are based on data from panellists or population survey, they are merely making concrete the value judgements of those groups. Furthermore, the use of a standardized system of measurement implies that one set of values is equally applicable to all individuals in the population which is itself a value-loaded assumption (Smith, 1987). In health measurement, the best that can be achieved is to make the value judgements used explicit so that those with other value systems can interpret the data.

References

Bergner, M., Bobbitt, R.A., Carter, W.B. & Gilson, B.S. (1981). The sickness impact profile: development and final revision of a health status measure. *Medical Care*, **19**, 787–805.

Capewell, G. (1988). Techniques of health status measurement using a health index. In *Measuring Health – A Practical Approach*. Teeling-Smith, G., ed., London: John Wiley.

Catford, J.C. (1983). Positive health indicators – towards a new information base for health promotion. *Community Medicine*, **5**, 125–32.

Coronary Prevention Group (1986). *Coronary Heart Disease in Asians in Britain*. London: Confederation of Indian Organisations.

Crawford, R. (1980). Healthism and the medicalisation of everyday life. *International Journal of Health Services*, **10**, 365–88.

Deyo, R.A., Inui, T.S., Leninger, J.D. & Overman, S.S. (1983). Measuring functional outcomes in chronic disease: a comparison of traditional scales and a self-administered health status questionnaire in patients with rheumatoid arthritis. *Medical Care*, **21**, 180–92.

Flemming, D.M., Crombie, D.L., Mayon-White, R.T. & Fowler, G.H. (1988). Comparison between the weekly returns of the Oxford Regional sentinel practice scheme for monitoring communicable disease. *Journal of the Royal College of General Practitioners*, **38**, 461–4.

Goldberg, D.P. & Hillier, V.F. (1979). A scaled version of the general health questionnaire. *Psychological Medicine*, **9**, 139–45.

Grogono, A.W. & Woodgate, D.J. (1971). Index for measuring health. *Lancet*, **ii**, 1024–6.

Hunt, S.M. (1986). Cross cultural issues in the use of socio-medical indicators. *Health Policy*, **6**, 149–58.

Hunt, S.M., McEwan, J. & McKenna, S.P. (1986). *Measuring Health Status*. London: Croom Helm.

Ibrahim, S. (1990). Measurement of impairment, disability and handicap. In *Measuring the Outcomes of Medical Care*, Hopkins, A. and Costain, D. eds, London: Royal College of Physicians of London.

Kind, P. (1982). A comparison of two models for scaling indicators. *International Journal of Epidemiology*, **11**, 271–5.

Latcham, R.W. & Kreitman, N. (1984). Regional variation in alcoholism rates in Britain: The effect of provision and use of services. *International Journal of Epidemiology*, **13**, 442–6.

Mahoney, F.I. & Barthel, D.W. (1965). Functional evaluation: the Barthel index. *Maryland State Medical Journal*, **14**, 61–5.

Martin, J., Meltzer, H. & Elliot, D. (1988). The prevalence of disability among adults. *OPCS Surveys of Disability in Great Britain Report 1*. London: HMSO.

McDowell, I. & Newell, C. (1987). *Measuring Health: A Guide to Rating Scales and Questionnaires*. New York: Oxford University Press.

McDowell, I. & Praught, E. (1982). On the measurement of happiness: an examination of the Bradburn scale in the Canada Health Survey. *American Journal of Epidemiology*, **116**, 949–58.

McKenna, S.P., McEwen, J., Hunt, S.M. & Papp, E. (1984). Changes in the perceived health of patients recovering from fractures. *Public Health*, **98**, 97–102.

Patrick, D.L., Sittampalam, Y., Somerville, S.M., Carter, W.B. & Bergner, M. (1985). A cross cultural comparison of health status values. *American Journal of Public Health*, **75**, 1402–7.

RCGP, OPCS & DHSS (1986). (Royal College of General Practitioners, Office of Population Census and Surveys, Department of Health and Social Security) Morbidity statistics from general practice 1981–82. *Third National Study Series MB5*, No. 1. London: HMSO.

Romeder, J.M. & McWhinnie, J.R. (1977). Potential years of life lost between ages 1 and 70: An indicator of premature mortality for health planning. *International Journal of Epidemiology*, **6**, 143–51.

Rosenberg, M. (1965). *Society and the Adolescent Self Image*. Princeton NJ: Princeton University Press.

Rosser, R.M. & Kind, P. (1978). A scale of valuation of states of illness – Is there a social consensus? *International Journal of Epidemiology*, **7**, 359–61.
Seedhouse, D. (1986). *Health: The Foundations for Achievement*. Chichester: John Wiley & Sons.
Shaper, A.G., Pocock, S.J., Phillips, A.N. & Walker, M. (1986). Identifying men at risk of heart attacks: strategy for use in general practice. *British Medical Journal*, **293**, 475–9.
Smith, A. (1987). Qualms about QALYs. *Lancet*, **i**, 1134–6.
Spitz, P.W. & Fries, J.F. (1987). The present and future of comprehensive outcome measures for rheumatic diseases. *Clinical Rheumatology*, **6**, Suppl. 2, 105–11.
WHO (World Health Organization) (1946). *Constitution*. Geneva: WHO.

5 *Developments in the assessment of physical activity*

PETER S.W. DAVIES

Introduction

Numerous methods have been used to assess levels of physical activity in both individuals and populations. All such techniques have advantages and disadvantages and none is entirely satisfactory. The importance of being able to monitor and measure activity levels can be shown easily (Malina, Panter-Brick, 1992; Hardman, this volume), and therefore new methods for measuring physical activity precisely and accurately are being considered constantly. Long-standing methods for the assessment of activity levels include those using mechanical devices, such as actometers, accelerometers and pedometers, heart rate monitoring, video recording and diary records. Some of these techniques do not necessarily measure all types of activity, i.e. pedometers and actometers, whilst others are not applicable to large groups of individuals, e.g. video recording, and others can be very time-consuming, e.g. diary records.

Physical activity is one component of total energy expenditure (TEE). By far the largest contribution to total energy expenditure is basal metabolic rate (BMR). The difference between BMR and TEE is the energy cost of thermoregulation, dietary induced thermogenesis, the energy cost of growth and energy expended in physical activity. The first two of these components are really quite small and, after the first 2 years of life, the energy cost of growth has diminished to a negligible amount. Thus the difference between TEE and BMR in children, adolescents and adults is primarily due to the energy expended in physical activity. The ratio of TEE to BMR or resting metabolic rate (RMR) is used frequently as an index of physical activity (e.g. Saris *et al.*, 1989; Davies *et al.*, 1991). This ratio is often known as the physical activity level (PAL).

Multiples of BMR or PAL values were used to determine energy requirements in adolescents by the FAO/WHO/UNU (1985) expert committee. Physical activity was divided into three gross categories: going to school and light activity, moderate activity and high activity. The energy expended in these activities as a multiple of BMR for boys and girls are shown in Table 5.1. The total daily PAL values calculated using these

Table 5.1. *The gross energy cost of activities defined as multiples of basal metabolic rate (BMR) (from FAO/WHO/UNU (1985) report)*

	Gross energy cost	
	Boys	Girls
Going to school and light activity	1.6 × BMR	1.5 × BMR
Moderate activity	2.5 × BMR	2.2 × BMR
High activity	6.0 × BMR	6.0 × BMR

multiples of BMR, and the time spent carrying out each of the activities varies over a rather narrow range from about 1.60 to 1.75 in boys and from about 1.50 to 1.65 in girls. The recent Department of Health (1991) report calculated that a 'non-active' individual would have a PAL value of 1.5 × BMR, whilst a very active individual spending 2 hours carrying out activity at 4 × BMR, e.g. walking on the level at an average pace, would have an overall PAL of 1.9.

Basal metabolic rate (BMR) or RMR can be measured with relative ease during childhood using indirect calorimetry (Benedict, 1914; Butte, O'Brian Smith & Garza, 1990). However, measurement of total energy expenditure in infants, children and adolescents was difficult, if not impossible, until the advent and subsequent development of the doubly labelled water techniques. Now estimates of physical activity via the TEE: RMR ratio are possible in the paediatric population.

The doubly labelled water technique

The doubly labelled water method was originally developed by Lifson and co-workers for use in small mammals. The principle of the method (illustrated in Figure 5.1) is that two stable isotopes of water ($H_2{}^{18}O$ and ^2H_2O) are administered simultaneously, and their initial enrichment in a body fluid (e.g. urine or saliva) and subsequent disappearance rates from this fluid are monitored by isotope ratio mass spectrometry. The initial increase in enrichment of either isotope reflects body-water pool size (from the principle of dilution) and permits an estimation of body composition. Subsequently, the disappearance rate of ^2H_2O reflects water output (and hence water intake); that of $H_2{}^{18}O$ reflects water output plus CO_2 production because ^{18}O is free to interchange between water and CO_2 through the action of carbonic anhydrase. The difference between the two

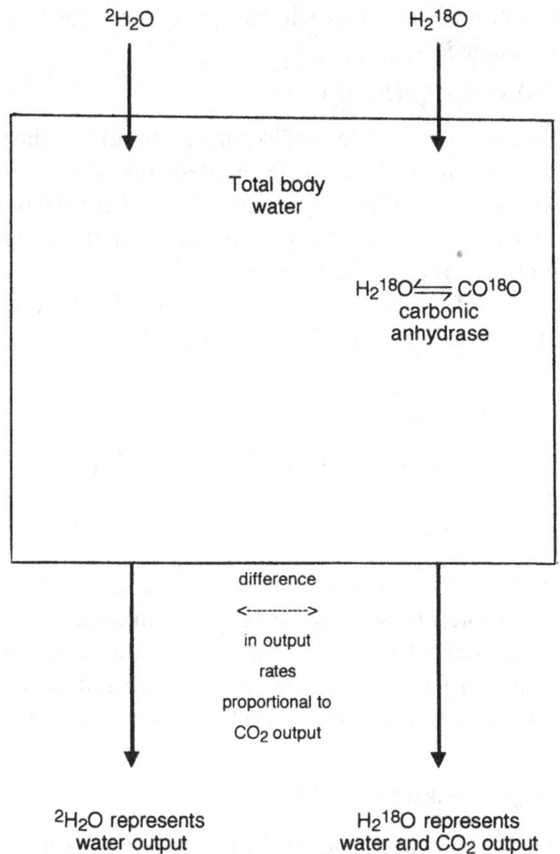

Figure 5.1. The principle of the doubly labelled water technique.

disappearance rates is therefore a measure of CO_2 production. Dosing can be carried out in the home, with daily urine samples being collected for 7, 10 or 14 days, depending upon the age of the child.

Isotopic enrichment of the urine samples is measured relative to a standard by isotope ratio mass spectrometry. Results are usually expressed as ‰ (per million) enrichment relative to a standard, thus:

$$\text{‰ enrichment} = ((R_S - R_R)/R_R) \times 10^3,$$

where R_S and R_R are the isotope ratios of the sample and reference water respectively. Linear regression equations are then obtained from the log transformed enrichments of ^2H and ^{18}O versus time. The regression coefficients are the disappearance rates, k_d and k_o, of ^2H and ^{18}O

respectively. The dilution space of each isotope at the beginning of the study period is calculated as

$$N=(TA/a)\times(E_a-E_t)/(E_s-E_p)$$

where N is dilution space (g). A is dose of isotope given (g), a is the portion of the dose retained for mass spectrometer analysis (g). T is tap water in which the portion (a) is diluted (g). E_a is enrichment of the portion. E_t is enrichment of tap water. E_s is the antilog of the intercept of the regression line. E_p is enrichment of the pre-dose urine sample.

Crude output rates of water and carbon dioxide (r^1H_2O and r^1CO_2 expressed as g equivalent H_2O/day) are calculated as

$$r^1H_2O=N_dk_d$$
$$r^1CO_2+r^1H_2O=N_ok_o$$

where N_d and N_o are the dilution spaces of the ^2H_2O and $H_2{}^{18}O$ respectively. Therefore crude CO_2 output is

$$r^1CO_2=N_ok_o-N_dk_d$$

True output rates of carbon dioxide are calculated, taking into account the proportion of water subject to isotopic fractionation in childhood (x), and values for fractionation of ^2H and ^{18}O (f_1 and f_2) in water vapour and ^{18}O (f_3) in CO_2. The value of x in the present study was 0.13 and the values for f_1, f_2 and f_3 were 0.93, 0.99 and 1.035 respectively. True CO_2 output rate was calculated as

$$rCO_2=(N_ok_o-N_dk_d(1+(f_2-f_1)x)/2f_3$$

Assuming the respiratory quotient, total energy expenditure can be calculated using Weir's (1949) formula.

Propagation of error analysis for the method can be carried out as described by Cole, Franklin & Coward (1990). The doubly labelled water technique has been validated for use in infants in two studies (Roberts *et al.*, 1986; Jones *et al.*, 1987). The technique has been reported as having an accuracy of $\pm 5\%$ and a precision of about $\pm 5\%$ during childhood.

A combination of measurements or predictions of basal metabolic rate and assessments of total energy expenditure using the doubly labelled water technique has now been used in a number of studies to assess levels of physical activity.

Physical activity during infancy

Until very recently, information pertaining to the energy cost of activity during infancy was almost nil. Experts were forced to speculate thus: 'As a guess since there is no hard information, and by analogy with adults, we might take the cost of activity at 3 months at +0.2 BMR for 12 out of 24 hours and at 6 months as +0.4 BMR for 12 hours in the day' (Waterlow, 1989).

Lawrence *et al.* (1991) have recently attempted to calculate the energy cost activity in infants using a factorial approach. The assumptions were that basal metabolic rate was 55 kcal/kg per day and that energy expenditure of lying, sitting and standing activities were 1.2 ×, 1.5 × and 2 × BMR. These estimates lead to an average PAL of 1.43 at 12 months of age for UK children. In contrast, using the same approach, Gambian children would have a mean PAL value of 1.30. British infants were observed to play for 2–3 times longer than Gambian children and to spend between two and four times longer undertaking 'vigorous activities' such as crawling, walking and running.

A large number of measurements of total energy expenditure have been carried out in infants using the doubly labelled water technique, and so it is possible to assess activity levels in infants if such data are combined with a measure of basal metabolic rate. However, it is impossible to meet the requirements of a measurement of basal metabolic rate with an infant subject. Moreover, the thermic effect of food and the energy cost of growth is probably always contributing to 'basal' metabolism in young children. Nevertheless, it is possible to measure metabolic rate during sleep which probably represents the minimum expenditure of energy in infancy. Sleeping metabolic rate (SMR) is remarkably constant and consistent in the first year of life (Schofield, Schofield & James, 1985; Davies & Joughin, unpublished observations), at about 50 kcal/kg body weight per day. Total energy expenditure is much more variable (Davies, Ewing & Lucas, 1989). The energy expended during activity, calculated as total energy expenditure minus SMR, in recent studies are shown in Table 5.2. The increase in energy expended in activity with age is apparent especially at 6 and 9 months of age. The apparent reduction at 4 months may be a function of the reduced number of infants studied at this age point. This possibility is enhanced as the PAL value shows an increasing trend from 1.5 months onwards. Thus the PAL values calculated for infants using the doubly labelled water technique are considerably higher than those estimated by the factorial approach of Lawrence *et al.* (1991).

Physical activity during childhood

The dose of stable isotope given to an individual to measure total energy expenditure is body weight dependent. The cost of $H_2{}^{18}O$ is currently about $70 per gram. Thus a 40 kg child, for example, will cost approximately $400 to dose. Therefore, measurements of total energy expenditure in

Table 5.2. *The energy cost of activity (kcal/kg body weight per day) and physical activity level (PAL) value for infants*

Age (months)	N	Energy cost of activity (kcal/kg per day)	PAL
1[a]	20	15	1.40
1.5[b]	49	17	1.34
3[b]	50	21	1.42
4[a]	20	16	1.50
6[b]	37	29	1.58
9[b]	22	39	1.78

[a]Butte *et al.*, 1990.
[b]Davies *et al.*, 1989; Davies & Joughin, unpublished observations.

older children have been restricted for economic rather than for scientific reasons. Nevertheless, during childhood, PAL values between about 1.4 and 2.0 have been reported, using the doubly labelled water technique and measurements of resting metabolic rate. Saris *et al.* (1989) found mean PAL values of 1.95 in boys and 1.75 in girls aged 8 to 10 years. These high values, compared with other recent estimates (Department of Health, 1991), are supported by the data of Davies *et al.* (1991) summarized in Table 5.3.

Physical activity during adolescence

Activity levels during adolescence calculated via the doubly labelled water method are possibly the most contentious. Using traditional methods, a number of recent studies (Dietz & Gortmaker, 1985; Gortmaker, Dietz & Cheung, 1990) indicate that there is a marked scale reduction in energy expenditure during adolescence possibly caused by a significant reduction in levels of physical activity. Moreover, heart rate monitoring has also produced data that suggest low levels of energy expenditure in adolescents. Armstrong *et al.* (1990) studied 266 children aged 11 to 16 years. Heart rate was monitored on three successive days. Boys or girls had heart rates in excess of 139 beats per minute for only 5% of the time. This level is thought to be the intensity required to stress the cardiovascular system. Armstrong *et al.* (1990) concluded that British children have surprisingly low levels of habitual activity.

Table 5.3. *Mean physical activity levels (PAL) for children aged 3–10 years*

	Boys		Girls	
Age (yrs)	N	Mean PAL	N	Mean PAL
3–4[1]	13	1.52	18	1.44
5–6[2]	12	1.76	16	1.73
7–8[1]	10	1.82	15	1.92
8.1 (mean)[2]	—	—	10	1.71
9–10[1]	14	1.84	15	1.65
9.3 (mean)[2]	9	1.95	—	—

[1]Davies *et al.*, 1991.
[2]Saris *et al.*, 1989.

However, the limited published data on total energy expenditure and resting or basal metabolic rate during later childhood and adolescence would not support the contention that activity levels are low in this age group. Davies *et al.* (1991) have reported high levels of activity in adolescents. These data are shown in Figure 5.2 with the Department of Health (1991) PAL values for a non-active, moderately and highly active individual. Note that a 24 hour PAL level of 2.0 indicates that non-sleep expenditure must average 2.5 × BMR, since 8 hours of sleep will be 1.0 × BMR.

Use of PAL values as an assessment of physical activity in a clinical condition. The Prader–Willi syndrome

The Prader–Willi syndrome is characterized by hypotonia in the neonatal period, hypogonadism, short stature, small hands and feet, mental retardation of a variable degree and obesity. The obesity usually develops within the first 4 years of life and, as adolescents and young adults, weight for height can exceed 200% of reference data (Coplin, Hine & Gormican, 1976). The exact contributions of a raised energy intake, a reduced energy expenditure and decreased levels of physical activity to the development and maintenance of the obesity are unclear (Holm, 1981; Nardella, Sulzbacher & Worthington-Roberts, 1983; Cassidy, 1984; Schoeller *et al.*, 1988). Studies that might shed light on these problems have been hampered by the inability to measure some of the parameters of interest, notably,

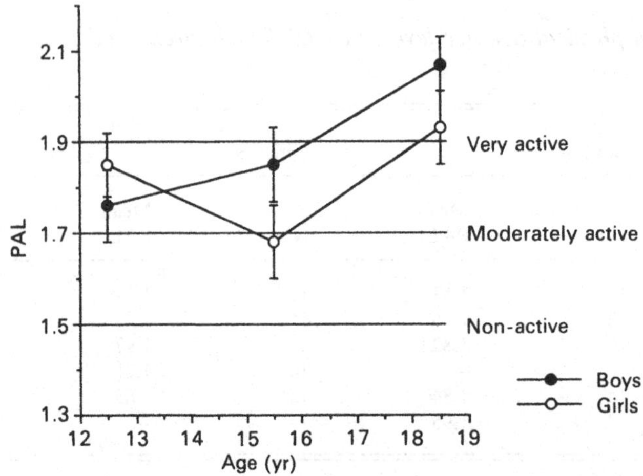

Figure 5.2. The mean (±SE) of PAL values for 12–18 year-old children plotted against the Department of Health (1991) values for non-active, moderately active and highly active individuals.

total energy expenditure and physical activity. Attempts have been made to assess activity levels using actometers and pedometers (Nardella *et al.*, 1983) but such techniques can be difficult to use in a handicapped population.

In order to assess levels of physical activity in this syndrome, ten subjects (five male, five female) were recruited with the assistance of the Prader–Willi Association (UK). All the subjects were living at home at the time of the study and eating a self-selected or parentally selected diet. Total energy expenditure was measured using the doubly labelled water technique as described above. Resting metabolic rate was measured using a portable open circuit indirect calorimeter. Subjects were required to rest for approximately 20 minutes before the measurement. Room air was drawn through a clear, plastic hood placed over the subject's head. Samples of gas leaving the hood were collected after 10 to 15 minutes. Ambient gas samples were collected simultaneously. Analysis of the carbon dioxide and oxygen concentration in the expired and ambient samples and knowledge of the flow rate through the hood allowed the calculation of resting energy expenditure.

The mean and standard deviation of a number of physical characteristics of the Prader–Willi children and the normal controls are shown in Table 5.4. The mean and standard deviation of total energy expenditure, resting metabolic rate and PAL value for the two groups are shown in Table 5.5.

Table 5.4. *The physical characteristics (mean and standard deviation (SD) of the Prader–Willi children and the normal controls*

	Controls		Prader–Willi	
	Mean	SD	Mean	SD
Age (yr)	12.56	4.2	11.65	3.5
Height (cm)	148.9	21.7	136.4	17.8
Weight (kg)	44.9	18.9	44.6	26.4
Fat free mass (kg)	34.1	15.1	23.9	10.3
% Fat mass	23.8	9.0	43.0	10.5

Table 5.5. *The mean and standard deviation (SD) of total energy expenditure, resting metabolic rate and PAL of the two groups*

	Control		Prader–Willi	
	Mean	SD	Mean	SD
Total energy expenditure (kcal/day)	2474	724	1758	569
Resting metabolic rate (kcal/day)	1615	377	1323	421
PAL value	1.53	0.23	1.33	0.21

After adjusting for age, fat free mass and sex by multiple regression analysis, total energy expenditure and resting metabolic rate were not significantly different between the Prader–Willi and control children ($t = -1.42$; $t = 0.19$ respectively). However, the TEE was reduced in comparison with the RMR. After adjusting for age, the PAL values of the Prader–Willi children were significantly lower than the control group ($t = 2.63$; $P < 0.05$), indicating that energy expended by physical activity in the Prader–Willi children was significantly reduced in comparison with normal children. Also the PAL values were negatively correlated with % body fat ($r = -0.42$; $P < 0.20$) and the sum of the two skinfolds ($r = -0.54$; $P < 0.10$). The development of obesity in children with this syndrome is usually associated with the marked hyperphagia exhibited in most cases. However, energy balance will be further compromised if energy expenditure is reduced due to reduction in activity.

Obese children have been reported as being less active in many studies

(Burch, 1940; Tolstrip, 1953; Gortmaker *et al.*, 1990), but whether the obesity makes activity difficult, or the lack of activity contributed to a positive energy balance and hence the development of obesity, is as yet uncertain. Only longitudinal studies of infants with Prader–Willi syndrome will lead to a greater understanding of the inter-relationships between energy intake and expenditure that leads to the obesity so often seen in the Prader–Willi syndrome.

A reduced level of total energy expenditure, probably due to a reduction in physical activity, in 12 week-old normal infants has been implicated in the later development of obesity at 1 year of age (Roberts *et al.*, 1988), although in a much larger study no such relationship was found (Davies, Day & Lucas, 1991). If activity levels could be increased in children with Prader–Willi syndrome, this may be an important contribution to the control of body weight and body composition.

This is an example where the doubly labelled water technique allows total energy expenditure to be measured non-invasively and thus does not impinge upon the individual's freedom to carry out his or her normal daily activity. Therefore it is likely that the technique itself will not cause changes in activity levels as can occur when using pedometers or other mechanical devices, and is of particular use in populations when classical techniques for the assessment of activity levels are inappropriate.

Summary

There are some data both from the UK and Holland to suggest that there may be a resurgence in physical activity amongst children, adolescents and young adults. Equally, the secular trend towards inactivity may not have affected these age groups as much as has been previously assumed.

However, there could be a number of methodological explanations for the finding of high activity. First the doubly labelled water technique must be assessed for any potential bias. There have now been many validation studies of the technique and none has shown any significant bias. This conclusion is supported by an expert working party report that examined all theoretical sources of error and bias (IDECG, 1990). A number of criticisms concerning the length of time over which post-dose samples should be collected, the representative nature or otherwise of measurements of total energy expenditure using the doubly labelled water technique and technical problems associated with the mass spectrometers most often used in the analysis of samples have recently been published (Durnin, 1990). However, many if not all of these supposed problems have long been recognized and have been discussed and overcome to the satisfaction of most of the IDECG Expert Committee (IDECG, 1990).

Secondly, it is possible that high levels of activity arose from biased sampling and that both the UK and Dutch data are based upon particularly active children. In both studies, the recruitment procedure was intended to be random but inevitably some self-selection may have occurred, with more active children volunteering for a study of this nature. Activity questionnaires in the UK children confirm that some of the children (notably some 7 year olds) were very active and participated in a number of sports but they were apparently not unusual and none of the subjects was in training for competitive sport. Thus it is unlikely that biased sampling is a major contributor to the high levels of physical activity.

References

Armstrong, N., Balding, J., Gentle, P. & Kirby, B. (1990). Patterns of physical activity among 11 to 16 year old British children. *British Medical Journal*, **301**, 203–5.

Benedict, F.G. (1914). The basal metabolism of boys from 1 to 13 years of age. *Proceedings of the National Academy of Sciences*, 7–10.

Burch, H. (1940). Obesity in childhood. *American Journal of Diseases in Childhood*, **60**, 1082.

Butte, N.F., O'Brian Smith, E. & Garza, C. (1990). Energy utilization of breast and formula fed infants. *American Journal of Clinical Nutrition*, **51**, 350–8.

Cassidy, S.B. (1984). Prader–Willi syndrome. *Current Problems in Pediatrics*, **14**, 1–55.

Cole, T.J., Franklin, M. & Coward, W.A. (1990). Estimates of error. In *The Doubly Labelled Water Method for Measuring Energy Expenditure. Technical Recommendations For Use in Humans*. Prentice, A.M., ed., Vienna: IAEA.

Coplin, S.S., Hine, J. & Gormican, A. (1976). Outpatient dietary management in the Prader–Willi Syndrome. *Journal of the American Dietitic Association*, **68**, 330–4.

Davies, P.S.W., Day, J.M.E. & Lucas, A. (1991). Early energy expenditure and later body fatness. *International Journal of Obesity*, **15**, 727–31.

Davies, P.S.W., Ewing, G. & Lucas, A. (1989). Energy expenditure in early infancy. *British Journal of Nutrition*, **62**, 621–9.

Davies, P.S.W., Livingstone, M.B.E., Prentice, A.M., Coward, W.A., Jagger, S.E., Stewart, C. & Whitehead, R.G. (1991). Total energy expenditure during childhood and adolescence. *Proceedings of the Nutrition Society*, **50**, 14A.

Department of Health (1991). Dietary reference values for food energy and nutrition for the United Kingdom. *Report on Health and Social Subjects*. No. 41. London: HMSO.

Dietz, W.H. & Gortmaker, S.L. (1985). Do we fatten our children at the TV set? Obesity and television viewing in children and adolescents. *Pediatrics*, **75**, 807–12.

Durnin, J.V.G.A. (1990). Methods to assess physical activity and the energy expended for it by infants and children. In *Activity, Energy Expenditure and Energy Requirements of Infants and Children*. Schurch, B. & Scrimshaw, N.S., eds, Lausanne: IDECG.

FAO/WHO/UNU (1985). Energy and protein requirements. *WHO Technical Report*, No. 724. Geneva: WHO.

Gortmaker, S.L., Dietz, W.H. & Cheung, L.W.Y. (1990). Inactivity, diet and the fattening of America. *Journal of the American Dietetic Association*, **90**, 1247–55.

Holm, V.A. (1981). The diagnosis of Prader–Willi syndrome. In *Prader–Willi Syndrome*. Holm, V.A., Sulzbacher, S. & Pipes, P.L., eds, pp. 27–44, Baltimore: University Park Press.

International Dietary Energy Consultancy Group (1990). *The Doubly Labelled Water Method for Measuring Energy Expenditure: Technical Recommendations For Use in Humans*. Prentice, A.M., ed., Vienna: IAEA.

Jones, P.J.H., Winthrop, A.L., Schoeller, D.A., Suyer, P.R., Smith, J., Filler, R.M. & Heim, T. (1987). Validation of doubly labelled water for assessing energy expenditure in infants. *Pediatric Research*, **21**, 242–6.

Lawrence, M., Lawrence, F., Durnin, J.V.G.A. & Whitehead, R.G. (1991). A comparison of physical activity in Gambian and UK children aged 6–18 months. *European Journal of Clinical Nutrition*, **45**, 243–52.

Nardella, M.T., Sulzbacher, S.I. & Worthington-Roberts, B.S. (1983). Activity levels of persons with Prader-Willi syndrome. *American Journal of Mental Deficiency*, **87**, 498–505.

Roberts, S.B., Coward, W.A., Schlingenseipen, K.H., Nohria, V. & Lucas, A. (1986). Comparison of the doubly labelled water ($^2H_2{}^{18}O$) method with indirect calorimetry and a nutrient balance study for simultaneous determination of energy expenditure, water intake and metabolizable energy intake in preterm infants. *American Journal of Clinical Nutrition*, **44**, 315–22.

Roberts, S.B., Savage, J., Coward, W.A., Chew, B. & Lucas, A. (1988). Energy expenditure and intake of infants born to lean and overweight mothers. *New England Journal of Medicine*, **318**, 461–6.

Saris, W.H.M., Emens, H.J.G., Groenenboom, D.C. & Westerterp, K.R. (1989). Discrepancy between FAO/WHO energy requirements and actual energy expenditure levels in healthy 7–11 year old children. *14th International Seminar of Paediatric Work Physiology*, Leuven, Belgium.

Schoeller, D.A., Levitsky, L.L., Bandini, L.G., Dietz, W.W. & Walczek, A. (1988). Energy expenditure and body composition in Prader-Willi syndrome. *Metabolism*, **37**, 115–20.

Schofield, W.N., Schofield, C. & James, W.P.T. (1985). Basal metabolic rate – review and prediction, together with an annotated bibliography of source material. *Human Nutrition: Clinical Nutrition*, **39C**, Suppl. 1, 5–41.

Tolstrip, K. (1953). On psychogenic obesity in children. *Acta Paedia*, **42**, 289.

Waterlow, J.C. (1989). Basic concepts in the determination of nutritional requirements of normal infants. In *Nutrition in Infancy*. Tsang, R.C. and Nichols, B.L., eds, Philadelphia: Hanley and Belfus Inc.

Weir, J.B. de V. (1949). New method for calculating metabolic rate with special reference to protein metabolism. *Journal of Physiology*, **109**, 1–9.

6 *Two national surveys of activity, fitness and health: the Allied Dunbar National Fitness Survey and the Welsh Heart Health Survey*

C.B. COOKE, G. DAVEY SMITH, G. HOINVILLE, J. CATFORD AND W. TUXWORTH

Introduction

This chapter presents information on two national surveys of activity, fitness and health: the Allied Dunbar National Fitness Survey and the Welsh Heart Health Survey. The former is ongoing, with data collection completed in November 1990, and the production of a report planned for early 1992. The latter was completed in 1985 under the auspices of Heartbeat Wales, a community-based heart disease prevention programme. While both are surveys of activity, fitness and health, they are different in many respects (e.g. aims and objectives, fieldwork procedures, test ingredients and methodology). The Allied Dunbar Fitness Survey is a large-scale descriptive survey specifically set up to address the measurement of physical activity patterns and fitness in a random sample of the adult population of England. In contrast, the major aim of the Welsh Heart Health Survey was that of intervention, attempting to change attitudes with a view to decreasing the incidence of coronary heart disease in the population of Wales.

Since the results of the Allied Dunbar Fitness Survey will not be available until 1992, only a description of the survey methodology and procedures is presented here. However, some of the results of the Welsh Heart Health Survey are presented to give an insight into the relationship between physical activity and fitness for a sample of the Welsh population. This gives an indication of the type of data and relationships that will be analysed in the Allied Dunbar National Fitness Survey, although the depth and breadth of activity and fitness data will be much greater than in the Welsh Heart Health Survey.

The Allied Dunbar National Fitness Survey

The role of the survey

The Allied Dunbar National Fitness Survey is a large-scale descriptive survey which focuses chiefly on the measurement of physical activity patterns and fitness levels of a random sample of the adult population in England.

The information, not previously available in this country, is needed as a data base to improve knowledge of the levels of physical activity and fitness amongst population subgroups and to improve understanding of the relationship between exercise, fitness and an individual's health and well-being, ultimately to provide new and valuable information for health education policies and exercise promotion.

An individual's health and well-being are dependent on a complex mix of factors many of which are beyond his or her direct control: genetic influences, the global environment and the way it is managed by society, local environmental factors such as pollution and overcrowding, resources available for health care and welfare provision, and so on.

Notwithstanding these community factors, there has been increasing attention during the twentieth century on the impact on health and well-being of an individual's own personal behaviour and lifestyle. The national survey of exercise, fitness and health fits within that context of personal behaviour and health care.

The focus on activity and fitness

Individual potential for physical fitness is genetically based. It is maintained and improved, however, by appropriate physical activity. The benefits of adopting a more active lifestyle, with an ensuing increase in fitness, are likely to be seen for all ages, in the easing of effort entailed in everyday activities, a lesser perception of fatigue, a bigger reserve capacity and a quicker recovery from extraordinary demands.

There is well established and increasing epidemiological evidence of a link between habitual physical activity and reduced risk of disease, and, more recently between physical fitness and reduced risk, particularly of coronary heart disease. For example, a recent American study of 4000 middle-aged men (Ekelund *et al.*, 1988) found that those who performed on a treadmill with only a modest increase in heart rate were less likely to die of coronary heart disease than those whose heart rates were more markedly elevated at the same level of exertion. This lower risk of heart attack appeared to operate independently of other major risk factors such as smoking, high blood pressure and high levels of cholesterol in the blood.

Another, larger American study (Blair *et al.*, 1989) of 10 000 men and 3000 women with no symptoms of disease concluded that people whose scores were low on a treadmill test were more likely to die of cardiovascular disease or cancer during the eight years of the study. Here, too, the reduced risk through fitness appeared to be independent of other risk factors.

An extensive review of 43 research studies which investigated the link between physical activity and coronary heart disease (CHD) was undertaken by the Center for Disease Control, Atlanta in 1987 (Powell *et al.*, 1987). That review found a consistent, statistically significant inverse association between physical activity and CHD in more than two-thirds of the studies.

The review also showed that the physical inactivity risk factor was independent of other key risk factors and was of similar magnitude to those risks (20 cigarettes or more per day, serum cholesterol greater than 268 mg dl^{-1}, and systolic blood pressure above 150 mm Hg).

The relative importance of the lack of physical activity as a risk factor for disease, and as a cause of incapacity and loss of functional ability, is greatly enhanced by the comparatively low levels of habitual physical activity among the adult population. For example, data collected by the Center for Disease Control, Atlanta, showed that approximately 60% of Americans do not perform physical activity with the regularity likely to offer protection against coronary heart disease (the criterion they used is a threshold of three or four times a week for at least 20 minutes) (Center for Disease Control, 1987).

In the United Kingdom, the most recently published paper (Morris *et al.*, 1990) shows that, among 9000 male civil servants, the rates of coronary heart disease over a 9-year period were significantly lower for those taking vigorous aerobic exercise than for the others in the survey.

The *New Case for Exercise* published jointly by the Sports Council and the Health Education Authority (Fentem, Bassey & Turnbull, 1988) identifies other potential health benefits of exercise, for example, in the treatment of diabetics, in protection against osteoporosis, and in reduction of hypertension.

Many functional abilities are reduced with ageing such as static and dynamic strength, muscular speed and power and maximal aerobic power. However, inactivity also leads to physiological changes which include poor cardiovascular function, weakness of muscles and reduced bone strength. An individual finds that, after a period of prolonged inactivity, he or she cannot work as hard or for so long; a greater effort is required than previously, and there is an increased sense of effort for any given task. Because levels of habitual physical activity among the general population are low, this disability, attributable to an inactive lifestyle, is an unnecessary but not uncommon cause of incapacity. The Toronto Conference concluded that 'Exercise training in the fifth and sixth decades of life may alleviate this decline, but more importantly will induce a functional gain, equivalent to as much as 10 to 15 years of ageing in many individuals' (Bouchard *et al.*, 1990).

Figure 6.1, which compares the four main CHD risk factors, demonstrates why health education to encourage exercise should be a priority. The other three risk factors have a much lower prevalence than the risk due to physical inactivity among the American population.

Information about the level of physical activity of the British population is sparse. In a pilot survey to develop the methodology for the national fitness survey (carried out in the West Midlands in 1988) results were similar to those found in Canadian and American studies. For example, the pilot survey results showed that:

- Only 6% of adults took part in regular weekly vigorous exercise.
- Around 30% engaged in some vigorous activity, but not regularly.
- Among people aged 60 and over, more than 50% did little activity of any kind.

The product of the survey

Epidemiological research relies heavily on the correct classification of people into subgroups (for example, distinguishing between those who exercise regularly, occasionally or never, or separating those with good cardiorespiratory performance from those with lower ability) in order to look at the association between variables, and to answer such questions as 'Do those who exercise regularly have higher cardiorespiratory performance than those who do not?'.

In order to collect data to look at the complex interaction between activity, fitness and health, considerable development work was necessary to produce comprehensive and accurate measurements of activity and fitness variables which could be applied to the general adult population. That programme of development work, at the Universities of Birmingham, London, Loughborough and Nottingham, funded by the Health Education Authority and the Sports Council, took place between 1985 and 1989.

Using the measurements developed during the preceding programme, the main survey will provide the first national audit of activity and fitness levels for different subgroups of the population of England. Both the measurements of activity and fitness are performed in greater depth than has previously been achieved in a large survey.

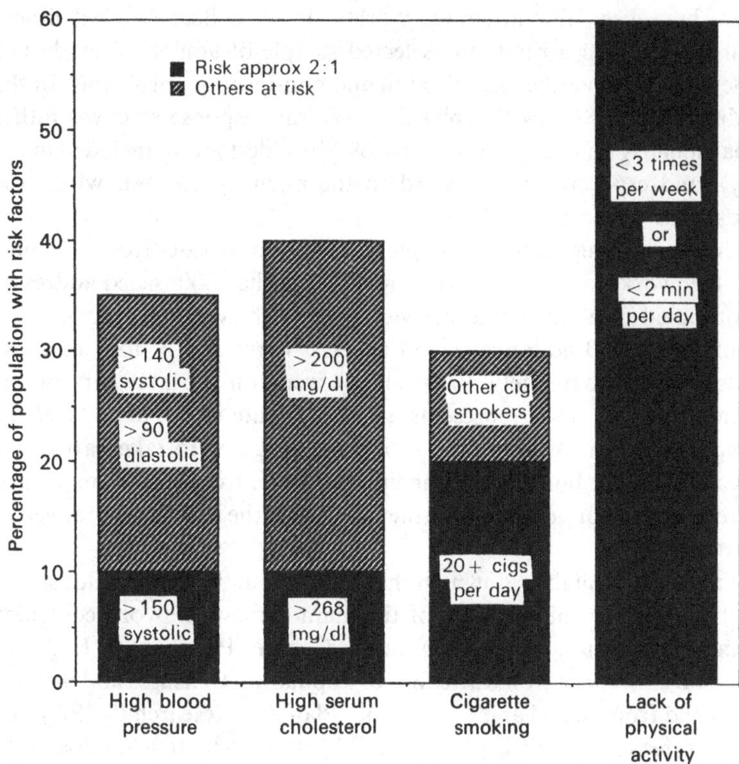

Figure 6.1. The four main coronary heart disease risk factors (Center for Disease Control, 1987).

The survey method

The Allied Dunbar National Fitness Survey consisted of interviews and fitness measurements of a large randomly selected sample of the adult population of England. Six thousand addresses were selected from the electoral registers, and one adult was chosen subsequently from each address.

Although the initial interviews took place at each person's home, the fitness measurements were carried out a few days later as a second stage, in a specially equipped Portakabin located at a nearby hospital site. This means that the sample is geographically clustered rather more heavily than would be the case for other interview surveys. Thirty parliamentary constituencies were randomly selected as the first stage sampling points and 200 addresses were then chosen, also at random, from within each constituency.

There have been few attempts, worldwide, to collect detailed fitness measurements from a randomly selected sample of adults; no study has completely overcome the self-selection and non-response problems. In the Canadian Fitness Survey (Shephard, 1981), the response rate was rather low, particularly among those aged 50–69 (the oldest group included in the survey), and especially with regard to the main fitness test, where the response was in the order of 35%.

At the time of preparing this chapter, only response rates from one third of our fieldwork are available. Around 1900 of the 2000 issued addresses were occupied dwellings, and interviews were achieved at approximately 71% of these 1900 addresses (1350 adults). Three out of four of those interviewed (approximately 1000 adults) went on to take part in the physical appraisal. That represents a response rate of about 53% of the occupied dwellings. A further 60 people (3%) agreed to take part in the physical appraisal, but, due to our tight schedule (resulting from limited resources) it was not possible to fit them in before the Portakabin moved to its next site.

The response is slightly down on that achieved in the earlier pilot study. Unfortunately, the initial phase of the main survey fieldwork coincided with the introduction of the Community Charge (Poll Tax) in England; there is some evidence from the Office of Population Censuses and Surveys (OPCS) and from Social and Community Planning Research (SCPR) that response rates for other surveys dropped by about 5% or so compared to the preceding period.

A response of 70% or more at the interview stage compares favourably with response rates to other large-scale complex national surveys such as the National Travel Survey, British Social Attitudes and the Cambridge Health and Lifestyle Survey. The response rate of slightly more than 55% at the fitness measurement stage is a little higher (closer to 60%) amongst those aged 16–74 years and then drops among the older age-groups. Fortunately, there will be a considerable amount of questionnaire data about activity levels, attitudes and self-assessment of fitness to identify whether, and in what ways, respondents at the second stage differ significantly from the non respondents at that stage. Possible biases created by this drop in response between the two stages will be identifiable.

Contact was then made at each selected address by an experienced interviewer from the Office of Population and Censuses and Surveys (OPCS) who selected one adult, using a specified random selection procedure, and arranged a time to interview that person. After completion of the questionnaire, the respondent was invited to attend a nearby Portakabin for a physical appraisal (predominantly fitness measurements). For subjects aged 75 years and over (and as an option for women aged 60 to

74), the fitness measurements were carried out at home, instead of at the Portakabin, on a subsequent visit by a member of the Portakabin fitness assessment team accompanied by an interviewer. This home appraisal excluded some of the muscle function measurements and the lung function measurement, all of which required bulky equipment, and the treadmill test, but included a few additional simple functional tests.

The survey had to be designed so that fieldwork could take place in several waves subject to available funds. The first wave covered 20 sampling points and took place between February and August 1990. The second wave extended the fieldwork to 30 sampling points, and was completed by mid-November 1990. A third wave was added in the form of an extension of the questionnaire part of the Survey only, which will also be completed by early 1992.

The interview

The home interview takes between $1\frac{1}{4}$ and $1\frac{1}{2}$ hours to complete, about half of which is concerned with questions about physical activity behaviour. This includes activity in the home (DIY, gardening and housework), at work and for recreation.

In any analyses of physical activity, no one reference period will serve all purposes. The questionnaire makes use of differing periods according to the nature of the data to be collected.

In defining these reference periods the following were considered:

1. the assessment of current and habitual physical activity either of which may play a part in determining current fitness levels;
2. some form of lifetime participation details to give insight into an individual's previous behaviour and propensity to engage in physical activity;
3. information about the adolescent period. This is important physiologically and also behaviourally, since non-participants at this age rarely become participants later on, and other studies have also shown this to be a period of decline in physical activity.

The assessment of current physical activity attempted to identify frequency, duration and intensity of exercise in order that threshold classifications could be derived. For example, those subjects who meet a widely used general prescription of exercise believed to convey a health benefit (vigorous aerobic exercise, three times a week for at least 20 minutes) can be identified. By relating physical activity to the fitness data, the threshold classification can be taken a step further and subjects categorized, also, on the basis of relative levels of physical exertion (i.e. their activity level in relation to their maximum physiological ability).

This will enable a more personalized approach to the threshold classification to be adopted by deciding for each individual whether he or she is engaging in a physical activity of sufficient intensity to have a 'training effect'.

Other sections of the questionnaire cover attitudes to exercise, fitness and health, other lifestyle behaviours and a fairly detailed health profile. Where possible standard questions have been incorporated in order to facilitate comparisons with other studies.

The following is a list of the items included in the questionnaire:

1. sex, age, marital status, occupation;
2. exercise and sport, current and past, access to facilities and barriers to participation;
3. physical activity at work, in housework, DIY and gardening, and in moving about (walking, cycling, stair-climbing);
4. other lifestyle and health-related behaviours (smoking, alcohol, dietary habits, etc);
5. health history and status (including screening for the physical appraisal);
6. sport-related injuries;
7. knowledge about exercise, attitudes towards exercise and health;
8. relevant psychological variables including well-being and social support, locus of control, stress and anxiety.

Medical screening

Reviewing the experience of the Canadian Home Fitness Test (Shephard, 1981), Shephard stated that the test had been used on about 500 000 subjects, of whom 50 000 were taking part in monitored tests, and that there had not been a single serious incident. High-risk subjects were excluded by the Physical Activity Readiness Questionnaire (PAR-Q), a screening questionnaire (Chisholm, Collis & Kulak, 1975).

The Welsh Heart Health Survey (Welsh Health Programme Directorate, 1987), discussed later in this chapter, used a test based on heart rate response to submaximal work on a cycle ergometer and was also completed without incident.

Elsewhere, Shephard has stated that exercise-related deaths in healthy adults are so rare that precise calculations of risk are almost impossible (Shephard, 1974). He estimated that, if 5 million middle-aged men and women each performed a 6-minute submaximal exercise test once per year (e.g. the Canadian Home Fitness Step Test) there would be one single 'critical' event every 16.5 to 33 years. Other data (Ellestad, 1980) suggest that a maximal test has a mortality rate varying between 1.7 per 10 000 and 2 per 100 000. But it should be noted that the exercise procedure used in the Allied Dunbar National Fitness Survey is not a maximal test. Exercise of this lower intensity is used in some medical centres within a week or two after an acute myocardial infarction.

For safety reasons, the Canadian Fitness Survey screened out many subjects from their physical appraisal. For example, 38% of those aged 50 to 59 years and 52% of those aged 60 to 69, all of whom had agreed to take part, were screened out. Overall, for all ages, the figure was 15%. The design for the survey has the advantage that people are brought to central laboratories which allows medical supervision to be provided for those at risk. A panel of doctors was recruited locally to assist in the screening of subjects prior to their fitness assessment and to observe those most at risk during the measurements. For all subjects on all occasions, medical safety was protected by the quality of care offered by the assessors (all of whom are trained and accredited in cardiopulmonary resuscitation and ECG recognition) and the availability of the hospital emergency services.

The physical appraisal

The physical appraisal consists of a battery of tests, selected to represent several of the facets that contribute to an individual's overall fitness. The tests and their reasons for inclusion are listed below.

Anthropometry:

Height, weight, demispan, skinfold thickness, waist and hip girths.

The body mass index, skinfolds and the waist and hip girths give important data regarding body fat and its distribution which are related to health and risk factors for disease. Measurement of body dimensions and composition is also necessary for interpreting other capacities.

Arterial blood pressure:

Blood pressure is measured because of the importance of hypertension as a cardiovascular risk factor and as a contra-indication for exercise testing.

Muscle function:

Isometric hand-grip strength, isometric leg extension strength and explosive power of the lower limb.

Strength and muscle power generally facilitate and are developed by participation in vigorous activity. More specifically, hand-grip strength is important in many aspects of daily life.

The quadriceps muscle extends the leg and is one of the most powerful in the body. Lack of strength in these muscles will affect most weight-bearing activities and general locomotion.

Explosive power of the lower limb, measured as the rate at which the muscles can extend the lower limb, enables performance of tasks such as rising from the seated position, take-off of normal and hurried strides, negotiating kerbs and steps, using stairs and especially avoiding tripping and falling.

Shoulder abduction

During and beyond middle-age, joints in old people become stiff and their range of motion is reduced. Shoulder abduction is an important movement involved in reaching above the head; this is necessary in daily life in, and outside, the home (e.g. on public transport) and in many sports and leisure-time activities.

Lung function

Simple ventilatory tests are included to assist in the interpretation of the results of other tests and to identify those with pulmonary disease contributing to disability. Forced vital capacity (FVC) and forced expiratory volume in one second (FEV1) are measured using a spirometer, and the ratio of FEV1/FVC analysed.

Cardiorespiratory exercise test

The capacity of the cardiorespiratory system is a central aspect of fitness relating to health and to participation in vigorous leisure pursuits. It determines the level of physical work which can be sustained and thus which of the activities of daily living are comfortably achieved or even possible. It reflects the extent to which an inactive lifestyle may be contributing to limitation and disability, on the one hand, and how far the physiological changes promoted by regular exercise may be conferring health benefits on the other. It is a complex attribute and, in terms of physical performance, has a profile specific to the activity for which it is required. It is influenced by both genetic endowment and adaptation through training.

The measurement of the cardiorespiratory response to exercise takes about 30 to 40 minutes to complete, much of which is taken up with explanations to the subject, a trial walk on the treadmill, placing ECG electrodes and familiarizing subjects to breathing through a mouthpiece. The submaximal test (which takes between 5 and 16 minutes) is based upon walking on a motorized treadmill at speeds not exceeding 5 km per hour (just under 3 miles per hour) with the gradient increasing by 2.5% per minute from the 6th to the 14th minute, and until the subject's heart rate has risen to his or her target heart rate adjusted for age. An alternative, less severe protocol has been devised for use by subjects with a very low level of fitness: it involves a constant low treadmill speed (3 km per hour) with smaller increments in gradient (2% per minute).

The condition of the subject and the subject's perception of the intensity of the exercise are monitored using the Borg rating scale of perceived exertion. Expired gas composition, ventilatory rate and volumes are metered continuously throughout the test, providing a measure of energy expenditure at each stage.

The cardiorespiratory measurements used in this survey will describe functional capacity well. It will be possible to determine, for each individual, conventional estimates of maximal oxygen uptake ($\dot{V}O_2$ max), derived from direct determination of $\dot{V}O_2$ during submaximal levels of exertion, but data will be available also for the derivation of many alternative indices. The exercise mode involves carrying the body weight, differs little from normal walking and is thereby relevant to the activities of daily living.

Summary

This completes the description of the methodology and field work procedures used in the Allied Dunbar National Fitness Survey. Data from the questionnaire, (of which 50% (45 minutes) is about activity), combined with the sophisticated fitness measurements carried out under laboratory conditions, provide the most detailed analysis of these variables to date. The survey should therefore provide valuable objective data on the fitness, activity and health status of a representative sample of the population of England.

Fitness and activity measures in the Welsh Heart Health Survey

The Welsh Heart Health Survey 1985 was carried out under the auspices of Heartbeat Wales, a community based heart disease prevention programme. The survey was designed for the dual purposes of 1) providing information regarding health status, health beliefs and health-related behaviours for planning of health promotion programmes; and 2) supplying baseline data for assessing the effects of the interventions. The detailed rationale for the survey design, the protocol and questionnaire and the field manual for the clinical procedures are all available (Catford, Nutbeam & Davey-Smith, 1985; Davey-Smith, 1986).

The study consisted of two component parts. The whole sample participated in a questionnaire survey (referred to as the community survey), and a subsample were invited for a screening examination of cardiovascular risk factors (referred to as the clinical survey (Davey-Smith 1991)). Households were sampled, with all members of these households between the ages of 12 and 64 included in the study.

Background and methodology

Sample selection, response and source of bias

The sample selection process was based on a three-stage procedure using health authorities as first-stage sampling points, with electoral registers forming the basic sampling frame. Contact was made with one or more residents in 18 921 of the 21 046 sample households. 32 461 people in these households constituted the sample for the survey and were left questionnaire booklets for self-completion and return. 21 637 (67%) usable questionnaires were returned.

Individuals were included in the total population for the clinical survey provided they returned signed consent forms with their questionnaires. In total 13 490 returned these consent forms representing 62% of responders to the questionnaire survey and 41.5% of the originally selected sample. From this group 2150 were randomly selected for clinical examination. Those attending for examination numbered 1832, representing a response rate of 85% of those selected.

Groups under-represented among respondents to the community survey included men under 25, men over 45 and the manual social classes of both sexes. The smoking status of the original household interview sample and the responders were similar, however, 32.8% of the original sample and 34% of the responders were reported as smokers.

The overall response rate to the clinical survey was low, at 36%. A major reason for this was the 'opt-in' nature of the sample, a consequence of asking subjects in the questionnaire if they would be willing to undergo clinical examination.

There is clearly the potential for response bias at all levels of the sampling procedure used in this study. However, data exist on age, sex, social class and smoking behaviour of the original household interview sample, and on a wide range of variables on the responders to the questionnaire survey. This allows a comparison of the clinical survey sample with the original sample, and with the questionnaire sample, on a wide range of demographic and behavioural variables.

There was a considerable social class bias in the clinical sample. Therefore, the clinical sample was compared to the community sample on factors related to social class. Whilst ex-smokers were over-represented in the clinical sample, the percentage of current smokers was similar in both. Of particular relevance is that there was little difference in reported participation in leisure time exercise by the two samples. Many other behavioural variables were examined, and the most marked difference remained that shown by smoking habits. Further details of the sample selection for this survey and of a study of non-responders are available (Davey-Smith & Nutbeam, 1990; Davey-Smith, 1991).

Clinical procedures

The following measurements relevant to the current report were made:

1. height and weight;
2. skinfold thicknesses measured at the biceps, triceps, sub-scapular and supra-iliac sites. Percentage body fat was estimated from these measurements using the Durnin and Womersley (1974) formulae;
3. waist and hip circumferences.

A manual describing the detailed procedures for all clinical measurements is available (Davey-Smith, 1986).

Suitability for fitness test

Suitability for the fitness test was determined by screening questionnaire, interview, blood pressure measurement and, if necessary a trial of using the ergometer. Before attending, the clinic subjects completed an adapted version of the Physical Activity Readiness Questionnaire (PAR-Q) (Chisholm *et al.*, 1975). This asks about possible reasons for exclusion: physical disability, chest pain and diagnosed heart disease or hypertension.

The completion of the adapted PAR-Q was followed by interview of all subjects by the doctor in attendance. Subjects feeling they had a disability which might render performing the test difficult were asked to undergo a brief trial on the ergometer.

Exclusion criteria were as follows:

1. subjects experiencing chest pain on walking at normal pace up a slight incline;
2. subjects taking betablockers;
3. subjects with a physical disability rendering cycling impossible;
4. subjects with an average recorded diastolic blood pressure above 110 millimetres of mercury or an average recorded systolic blood pressure above 170 millimetres of mercury

According to the above criteria, 149 subjects (77 males and 72 females) were unable to perform a fitness test.

Cycle ergometer protocol

The basic test mode was one that had been used in a large survey of male factory workers (Tuxworth *et al.*, 1986) and in the 'Eurofit' test on several thousand children (Council of Europe, 1983). The test suffers from the limitations of submaximal tests (Astrand & Rodahl, 1986). Test re-test reliability for the test used here has been excellent ($r=0.977$ between test occasions 1 and 3) for physical education students in the familiar surroundings of their laboratory, and only moderate ($r=0.83$) for 8 year-old children with a 'stranger' administering the test (data of Tuxworth and Cooke, personal communication).

The bicycle ergometer test is a sub-maximal progressive and continuous test usually consisting of three times 3 minute workloads. The test is based on the measurement of heart rate response to three individually tailored work intensities. Attainment of similar relative levels of exertion by all subjects is facilitated by the use of the heart rate response to the previously completed workload in the setting of the following workload. The lower the heart rate response to a given workload, the greater the subsequent workload increment, and vice versa.

The aim of the protocol is to elevate heart rate response to approximately 80–85% of predicted age-related maximal heart rate, estimated by 85% of [220 – age] based on recommended heart rates for the cessation of sub-maximal exercise tests in heterogeneous populations (Astrand, 1967). The initial workload can be set on the basis of either 1) lean body mass or 2) body mass (if percentage body fat is not estimated via skinfold measurement). Workload increments are based on the heart rate response to the previously completed workload.

If the subject's heart rate response to workload 3 has not reached a value within 15 beats per minute of 85% at predicted age-related maximal heart rate then a fourth workload is administered.

The procedure for incrementing workload was derived from the heart rate/workload tables published in the provisional Eurofit manual (Council of Europe, 1983).

To ease the burden of administration, a computer program was developed which provided the tailored workload increments in response to the input of the heart rate for each completed workload. Indices of physical work capacity are calculated by linear extrapolation or interpolation and a computer solution of the Astrand-Ryhming (Astrand & Ryhming, 1954; Shephard, 1970) nomogram was used for the prediction of $\dot{V}O_2max$.

Safety features: If the heart rate response exceeded 85% of the predicted age-related maximal heart rate and this was sustained for 15 seconds at any time during the test, the workload was decreased immediately and the subject pedalled for a further short period at a loading similar to the initial load for the purpose of cooling down.

Fitness test results

Predicted maximal oxygen uptake and physical work capacity
Prediction of maximal oxygen uptake by the Astrand and Ryhming nomogram required subjects to complete at least one workload and for heart rate to be elevated into the linear range for extrapolation (122 to 170 for men; 120 to 172 for women). Predicted $\dot{V}O_2$max results were obtained on 1643 (90%) of the 1832 subjects attending for clinical examination.

Table 6.1 presents predicted maximal oxygen uptake results by age and sex, Table 6.2 giving weight-related scores ($\dot{V}O_2$max/weight). Predicted $\dot{V}O_2$max is higher in males than females. $\dot{V}O_2$max and $\dot{V}O_2$max/weight fall steadily with age, the rate of decline being greater in males than females. The percentage difference between male and female $\dot{V}O_2$max and $\dot{V}O_2$max/weight therefore falls with advancing years. For $\dot{V}O_2$max/weight, the female levels are 72% of male levels in the youngest age group, this percentage steadily increasing to 91% in the oldest age group.

The prediction of physical work capacity (watts) requires subjects to complete at least two workloads. For extrapolation to a heart rate of 170 beats per minute, or its age-related equivalent, the heart rates should be reasonably spread throughout the linear range and the upper heart rate should be relatively close to the criterion heart rate (i.e. 85% predicted age-related maximal heart rate). Physical work capacity results were obtained on 1583 (86%) of the 1832 subjects attending for clinical examination.

Physical work capacity falls steadily with age. Male scores are greater than female scores, although unlike in the case of predicted maximal oxygen uptake the female to male ratio remains constant across the age groups, being approximately 60% for age-related physical work capacity and 70% for age and weight-related physical work capacity.

Body dimensions, body composition and fitness measures
The strong relationships between fitness, body dimensions and indices of body composition confound the associations between fitness and other variables. Since body composition changes rapidly around puberty, only subjects aged 18 and over are considered, and all further analyses are restricted to this age-group.

The correlations between the fitness indices and age, weight, height, percentage body fat (% body fat), lean body mass (LBM), fat mass (FBM), body mass index (BMI) and waist to hip ratio (W/H) are given in Table 6.3. $\dot{V}O_2$max ($1\ min^{-1}$) is more strongly positively correlated with age in males than females. In both sexes $\dot{V}O_2$max is positively related to weight and height, this is stronger in the case of height. For men, $\dot{V}O_2$max is related to

Table 6.1. *Predicted maximal oxygen uptake by age and sex in the Welsh Heart Health Survey*

Age group (yr)	$\dot{V}O_2max$ ($l\ min^{-1}$)					
	Male			Female		
	Mean	N	SD	Mean	N	SD
12–17	3.03	74	0.86	2.01	64	0.54
18–24	3.30	84	0.79	2.02	101	0.57
25–29	3.11	79	0.76	1.93	69	0.48
30–34	3.07	80	0.79	1.88	92	0.43
35–39	2.97	100	0.69	1.81	101	0.45
40–44	2.72	116	0.71	1.83	93	0.55
45–49	2.58	87	0.62	1.69	71	0.48
50–54	2.36	84	0.63	1.82	80	0.62
55–59	2.26	65	0.54	1.69	86	0.56
60–65	2.32	62	0.73	1.74	55	0.66

SD = standard deviation.

Table 6.2. *Weight-related predicted maximal oxygen uptake; by age and sex*

Age group (yr)	$\dot{V}O_2max$/weight ($ml\ kg^{-1}\ min^{-1}$)					
	Male			Female		
	Mean	N	SD	Mean	N	SD
12–17	54.50	74	10.39	39.26	64	9.01
18–24	47.66	84	10.87	34.61	101	9.59
25–29	41.05	79	9.25	31.83	69	8.02
30–34	39.45	80	10.03	30.28	92	7.29
35–39	38.33	100	8.96	30.08	101	8.32
40–44	35.11	116	8.89	29.22	93	9.17
45–49	33.32	87	8.30	26.96	71	8.76
50–54	30.47	84	8.52	28.71	80	10.31
55–59	29.49	65	6.96	26.13	86	7.71
60–65	29.94	62	8.38	27.25	55	11.78

SD = standard deviation.

Table 6.3. *Correlations of fitness and body composition variables in adults (> 18 years) with significance levels and sample number*

	Males		Females	
	$\dot{V}O_2max$	$\dot{V}O_2max/WT$	$\dot{V}O_2max$	$\dot{V}O_2max/WT$
Age	0.43***	−0.51***	0.18***	−0.26***
	756	756	745	745
Weight	0.19***	−0.29***	0.19***	−0.34***
	756	756	745	745
Height	0.31***	0.08*	0.26***	0.05^{ns}
	756	756	745	745
BMI	0.02^{ns}	−0.37***	0.08*	−0.38***
	756	756	745	745
Body fat	0.14***	−0.46***	0.00^{ns}	−0.39***
	747	747	735	735
FBM	0.01^{ns}	−0.42***	0.12**	−0.38***
	747	747	735	735
LBM	0.33***	−0.09**	0.27***	−0.23***
	747	747	735	735
W/H	−0.16***	−0.37***	-0.03^{ns}	−0.21***
	749	749	732	732

BMI = Body mass index; FBM = Fat mass; LBM = Lean body mass; W/H = Waist to hip ratio.
*** = $P<0.001$; ** = $P<0.01$; * = $P<0.05$; ns = not significant.

LBM but not FBM. In men, but not in women, $\dot{V}O_2max$ is negatively related to waist to hip ratio.

The weight-related measures of maximal oxygen uptake and physical work capacity, i.e. the raw scores divided by weight, are more strongly negatively correlated with age than are the unadjusted values. These indices are negatively associated with weight, BMI, body fat percentage, LBM, FBM and waist to hip ratio. Such measures should not be referred to as 'weight-corrected', although they often are. The weight-related scores penalize subjects for being of large mass, even if their size reflects frame rather than obesity.

To allow for the strong influence of age on fitness and on measures of body composition partial correlation coefficients were calculated, controlling for age. However, relationships between fitness indices and measures of body composition remained substantial.

Regression of raw fitness scores against age, age and weight, and age LBM and FBM was undertaken to examine the contribution of these factors in explaining population variance of fitness scores. The proportions

Table 6.4

Independent variables	$\dot{V}O_2max$ Male	Female
Age	19%	4%
Age and weight	25%	10%
Age, LBM and FBM	34%	16%

of variance accounted for, as indexed by the square of the multiple correlation coefficients, are given in Table 6.4.

Two strategies for dealing with these potential confounding factors have been adopted. Simple associations, e.g. between smoking groups and fitness have firstly been reported with raw scores ($\dot{V}O_2max$ and PWC85%) and with weight-related scores, raw scores divided by weight. These approaches are the standard ones in the current fitness literature, although they lead to associations strongly confounded by age and body composition. Reporting results in this fashion has the advantage of allowing comparisons with other studies to be made, however.

To control for potential confounders in further analysis adjusted raw score means have been calculated, adjusted for age, LBM and FBM by multiple linear regression. These adjusted means are reported for group comparisons. When relating continuous variables to fitness, partial correlation coefficient analysis, that is, correlation of residuals following multiple linear regression against the independent variables age, LBM and FBM, has been used to control for age and body composition.

Physical fitness and physical activity

Work activity Subjects were asked to rate the degree of exertion required by their work or other day-time activity. Four categories were allowed, according to main activity; sitting; standing or walking; lifting lift loads and/or frequent stair climbing; heavy work.

Mean fitness scores adjusted for age, LBM and FBM were then calculated (Table 6.5). For males, the adjusted mean $\dot{V}O_2max$ and PWC85% scores were highest for the heavy work group, but only significantly so for the comparison with the lifting group. For women no differences were significant, although for both $\dot{V}O_2max$ and PWC85% the heavy work group had the highest adjusted mean, followed by the lifting group.

Table 6.5. *Fitness scores adjusted for age, lean body mass and fat mass by activity at work*

	$\dot{V}O_2max$ (l min^{-1})					
	Male			Female		
	Mean	N	SD	Mean	N	SD
Sitting	2.76	180	0.63	1.79	138	0.50
Standing	2.77	269	0.64	1.83	517	0.50
Lifting	2.70	132	0.63	1.89	68	0.50
Heavy work	2.85	154	0.63	1.97	10	0.47

SD = standard deviation.

Leisure time activity The number of times that subjects had participated for at least 20 minutes in physical activity during their leisure time in the week preceding questionnaire completion was recorded. Three levels of activity were set: light (walking, light gardening, etc), moderate (swimming, cycling, golf, heavy DIY, etc), and strenuous (rugby, competitive swimming or cycling, squash, etc).

Correlations were calculated between frequency of activity participation and fitness scores. For men, strenuous activity was most strongly related to all fitness measures, with moderate activity showing weaker associations and light activity being (non-significantly) negatively related. For women, a gradient was seen, with all fitness indices showing strongest correlations with strenuous activity, and weakest correlations with light activity. Partial correlations controlling for age, LBM and FBM (Table 6.6) were then calculated. The correlation coefficients between fitness and both moderate and strenuous activity become slightly smaller for men and women.

Subjects were classified into four activity groups according to the intensity of leisure time activity. Subjects reporting any strenuous activity were placed in the 'strenuous' category, subjects reporting no strenuous activity but some moderate activity entered the 'moderate' category. If subjects reported no strenuous or moderate activity but some light activity they were categorized as 'light', with subjects reporting no activity being labelled 'none'. Table 6.7 displays fitness scores by these categories. The expected trend is seen, with the exception of the 'none' category which consistently demonstrates higher fitness scores than the 'light' group. In all cases, the 'strenuous' group demonstrated significantly (at the $P<0.05$ level) higher fitness scores than the other groups. These group differences

Table 6.6. *Partial correlations of leisure time activity and fitness controlling for age, lean body mass and fat mass (correlation coefficients, significance levels and sample number)*

	$\dot{V}O_2$max residual	
	Male	Female
Light residual	0.06[ns]	0.09*
	738	727
Moderate residual	0.10*	0.06[ns]
	741	732
Strenuous residual	0.22***	0.11**
	742	731

*** = $P<0.001$; ** = $P<0.01$; * = $P<0.05$; ns = not significant.

Table 6.7. *Fitness measures by leisure time activity intensity group*

(*a*)	Male: $\dot{V}O_2$max (l min^{-1})		
	Mean	N	SD
Strenuous	3.33*	146	0.84
Moderate	2.76	281	0.67
Light	2.44	244	0.62
None	2.73	71	0.75

(*b*)	Female: $\dot{V}O_2$max (l min^{-1})		
	Mean	N	SD
Strenuous	2.19*	38	0.55
Moderate	1.87	210	0.43
Light	1.77	395	0.60
None	1.85	93	0.58

$P^* = <0.05$.
SD = standard deviation.

were further examined by computing means of $\dot{V}O_2$max and PCW85% adjusted for age, LBM and FBM (Figure 6.2). In women, $\dot{V}O_2$max and PWC85% were significantly higher in the most active group than all other groups, no other comparisons being significant.

Physical fitness and social class

Social class was coded according to the Registrar General's classification, with groups I, II and IIINM being grouped as 'non-manual', and groups IIIM, IV and V being grouped as 'manual'. There were 175 participants who were economically inactive. They were not assigned to a social class group. Fitness scores according to these groups are shown in Table 6.8. In general, non-manual class men tend to be fitter than manual class men whilst manual class women tend to be fitter than non-manual class women. Only the difference for $\dot{V}O_2$max in women reaches the $P<0.05$ significance level, however. The same relationships are seen, and are strengthened by, adjustment for age, LBM and FBM.

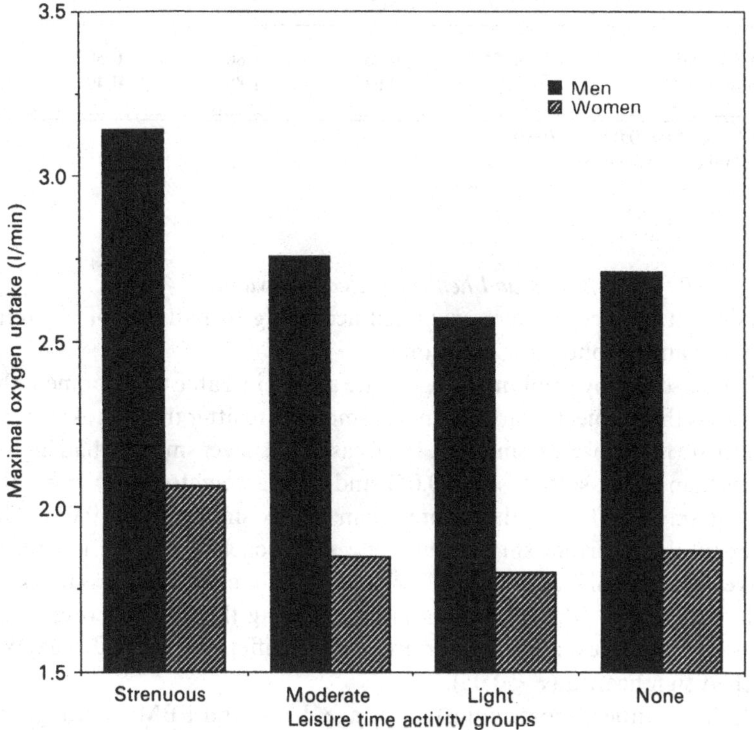

Figure 6.2. Maximal oxygen uptake scores adjusted for age, lean body mass and fat mass by leisure time activity intensity group.

Table 6.8. *Fitness according to social class group*

(*a*) Unadjusted means

	$\dot{V}O_2max$ (l min^{-1})					
	Male			Female		
	Mean	N	SD	Mean	N	SD
Non-manual	2.78	384	0.78	1.80*	443	0.42
Manual	2.74	365	0.76	1.89*	294	0.51

(*b*) Adjusted means (Adjusted for age, lean body mass and fat mass)

	$\dot{V}O_2max$ (l min^{-1})			
	Male		Female	
	Mean	SD	Mean	SD
Non-manual	2.81**	0.65	1.80	0.50
Manual	2.72	0.63	1.88*	0.49

*$P<0.05$; **$P=0.07$; ***$P=0.11$.
SD = standard deviation.

Physical fitness and health-related behaviours

Physical fitness scores were examined according to patterns of cigarette smoking and alcohol consumption.

Fitness scores by smoking category are shown in Table 6.9. For men, the pattern is that subjects who have never smoked are fitter than smokers who in turn are fitter than ex-smokers. In all cases, the never smokers had higher scores than the ex-smokers ($P<0.05$), and for the weight-related scores the current smokers had higher scores than the ex-smokers ($P<0.05$). The never smoker, current smoker differences approached but did not reach conventional significance levels. For women, differences were less marked, with the subjects who had never smoked having the lowest scores in all cases. Only the never smoke, current smoker difference for $\dot{V}O_2max/WT$ reached significance ($P<0.05$).

For men, fitness scores adjusted for age, LBM and FBM reduce group differences. The apparent advantage smokers have over ex-smokers disappears. Only the higher PWC85% scores in never smokers compared

Table 6.9. *Fitness by smoking behaviour*

(*a*) Unadjusted means

	$\dot{V}O_2max$ (l min^{-1})					
	Male			Female		
	Mean	N	SD	Mean	N	SD
Smoker	2.77	255	0.79	1.88	200	0.56
Ex-smoker	2.61	235	0.76	1.88	202	0.57
Never smoked	2.90	264	0.81	1.78	344	0.55

(*b*) Adjusted means (Adjusted for age, lean body mass and fat mass)

	$\dot{V}O_2max$ (l min^{-1})			
	Male		Female	
	Mean	SD	Mean	SD
Smoker	2.76	0.64	1.87	0.51
Ex-smoker	2.73	0.66	1.88	0.51
Never smoked	2.81	0.64	1.79	0.50

SD = standard deviation.

to smokers and ex-smokers now approach significance (both $P<0.09$). For females, never smokers have lower $\dot{V}O_2max$ scores than ex-smokers ($P<0.05$) and current smokers ($P<0.08$), but PWC85% scores are similar for all groups.

Fitness by categories of alcohol consumption are shown in Table 6.10. Subjects reporting less than weekly drinking were categorized as occasional drinkers. Subjects drinking more frequently than this reported the average number of units they had drunk per week over the past year, and were assigned to three groups on the basis of tertiles of this.

For both sexes, the tendency is for fitness to be lower in the none and occasional drinker categories than the regular drinker categories. The only difference for males reaching conventional levels of significance was the higher fitness of moderate drinkers compared to occasional drinkers. For women in all cases, the heavy drinkers had significantly higher fitness than the occasional drinkers.

Table 6.10. *Fitness by pattern of alcohol consumption*

(*a*)	Male: $\dot{V}O_2$max ($l\ min^{-1}$)		
	Mean	N	SD
None drinker	2.54	33	0.68
Occasional drinker	2.64	187	0.68
Light drinker	2.68	140	0.71
Moderate drinker	2.90	202	0.85
Heavy drinker	2.86	188	0.82

(*b*)	Female: $\dot{V}O_2$max ($l\ min^{-1}$)		
	Mean	N	SD
None drinker	1.88	65	0.56
Occasional drinker	1.77	294	0.51
Light drinker	1.81	153	0.49
Moderate drinker	1.81	117	0.53
Heavy drinker	2.02	114	0.64

SD = standard deviation.

Adjusting means for age, LBM and FBM left the pattern of the results largely unchanged. For men, the fitness advantage amongst drinkers was seen predominantly for the moderate drinkers, whilst for women it was the heaviest drinking category who showed the highest fitness levels.

Discussion

Before discussing the activity and fitness results from the Welsh Heart Health Survey, consideration must be given to the sampling procedure. The clinical sample was based on respondents to the questionnaire survey who were willing to attend for a clinical examination. Although data exist which allow a comparison of the clinical and questionnaire samples, doubts remain as to how representative the clinical sample is for the purposes of making estimates regarding population fitness levels in Wales. Nevertheless, the data from the Welsh survey is presently the most representative regarding fitness (and hence the relationship between fitness and other variables such as activity) for the UK. However, the results of the Allied Dunbar National Fitness survey should supersede it in this respect.

In line with some other studies, anomalous results were obtained in the Welsh survey regarding fitness and health-related behaviours. Female non-smokers were less fit than ex-smokers, and for both sexes alcohol consumption was positively associated with fitness. The literature provides studies in agreement with these findings (Gyntelberg & Meyer, 1974; Jonsson & Astrand, 1979) and studies reporting opposite effects (Peters *et al.*, 1983; Tuxworth *et al.*, 1986). Physiological theory and experimental studies would not predict a beneficial effect of these behaviours on fitness (Astrand & Rodahl, 1986), and the inconsistent reports suggest that smoking and drinking may be related differentially to exercise and fitness in the different populations, and this may account for the diverse findings. In Wales, the predominant sporting activity, rugby, is traditionally followed by beer drinking, which could account for the higher fitness of moderate drinkers. In the present data set, subjects who reported performing strenuous leisure time activity were the group with the highest percentage of heavy drinkers suggesting that it is possible that leisure time strenuous activity confounds the alcohol–fitness association.

Fitness is related in the expected way with strenuous leisure time activity, and the strength of the association is of the degree which would be expected from the known cardiovascular training effect of exercise. In line with most (DeBacker *et al.*, 1981; Wilhelmsen *et al.*, 1976; Gyntelberg, 1975), but not all (Jonsson & Astrand, 1979) studies, leisure time activity was more strongly related to fitness than occupational activity. In agreement with studies of the necessary intensity of exercise to produce a training effect (Wenger and Bell, 1986) strenuous exercise was the modality most strongly related to fitness, this being especially true for males.

The fitness test results could be viewed as validation of the simple leisure time activity questionnaire used in the Welsh survey. Comparing the strength of association reported for the Welsh survey with the studies reviewed by Lamb and Brodie (Lamb & Brodie, 1991) reveals a similar magnitude of association to those seen with other general population samples.

Given that they are the most representative data for the UK at the present time, the results from the Welsh Heart Health survey presented here show the potential and the need for detailed information on both the fitness and activity patterns and their inter-relationships in the UK population. The Allied Dunbar National Fitness Survey should provide this information for a more representative sample of the population of England, using both a more detailed activity questionnaire and a larger number of more sophisticated fitness indices.

References

Albrecht, K.L. (1978). The W 170, an evaluation of measuring methods for fitness on naval personnel. In *Proceedings of the Symposium on Physical Fitness with Special Reference to the Military*, NATO, ed., pp. 135–47, NATO.

Astrand, I. (1967). The Scandinavian Committee on ECG classification: the 'Minnesota Code' for ECG classification: adaptation to CR leads and modification of the code for ECGs recorded during and after exercise. *Acta Medica Scandinavica.*

Astrand, P.O. & Rodahl, K. (1986). *Textbook of Work Physiology*, New York: McGraw-Hill.

Astrand, P.O. & Ryhming, I. (1954). A nomogram for calculation of aerobic capacity (physical fitness) from pulse rate during submaximal work. *Journal of Applied Physiology*, **7**, 218–21.

Blair, S.N., Kannel, W.B., Kohl, H.W., Paffenbarger, R.S., Clark, D.G. & Cooper, K.H. (1989). Physical fitness and all cause mortality. *Journal of the American Medical Association*, **262**, 2395–405.

Bouchard, C., Shephard, R.J., Stephens, T., Sutton, J.R. & McPherson, B.D., eds, (1990). *Exercise, Fitness and Health.* Champaign, Illinois: Human Kinetics Books.

Catford, J.C., Nutbeam, D. & Davey-Smith, G. (1985). *Welsh Heart Health Survey, 1985: Protocol and Questionnaire*, Cardiff: Heartbeat Wales.

Center for Disease Control (1987). Protective effects of physical activity on coronary heart disease. *Morbidity and Mortality Weekly Report*, **36**, No. 26. Atlanta.

Chisholm, D.M., Collis, M.L. & Kulak, L.L. (1975). Physical activity readiness. *British Columbia Medical Journal*, **17**, 375–8.

Council of Europe (1983). Testing physical fitness. In *Eurofit: Experimental Battery Provisional Handbook*, Strasbourg: Council of Europe.

Davey-Smith, G. (1986). Welsh Heart Health Survey 1985. In *Clinical Survey Manual*, Cardiff: Heartbeat Wales.

Davey-Smith, G. (1991). Physical fitness and risk factors for coronary heart disease. MD Thesis, Cambridge University.

Davey-Smith, G. & Nutbeam, D. (1990). Assessing non-response bias: a case study from the 1985 Welsh Health Survey. *Health Education Research*, **5**, 381–6.

DeBacker, G., Kornitzer, M., Sobolski, J., Dramiax, M., Degre, S., De Marneffe, M. and Denolin, H. (1981). Physical activity and physical fitness levels of Belgian males aged 40–55 years. *Cardiology*, **67**, 110–28.

Durnin, J.V.G.A. & Womersley, J. (1974). Body fat assessed from total body density and its estimation from skinfold thickness: measurements on 481 men and women aged from 16 to 72 years. *British Journal of Nutrition*, **32**, 77–97.

Ellestad, M. (1980). *Stress Testing, Principles and Practice.* 2nd edn, Philadelphia: F.A. Davies Co.

Ekelund, L., Haskell, W.L., Johnson, J.L., Whaly, P.S., Criqui, M.H. & Sheps, D.S. (1988). Physical fitness as a predictor of cardiovascular mortality in asymptomatic North American men. *New England Journal of Medicine*, **319**, 1379–84.

Fentem, P., Bassey, E.J. & Turnbull, N. (1988). *The New Case for Exercise.* Health Education Authority and the Sports Council: London.

Gyntelberg, F. (1975). Physical fitness and coronary heart disease in Copenhagen men aged 40–59: III. *Danish Medical Bulletin*, **21**, 49–56.

Gyntelberg, F. & Meyer, J. (1974). Relationship between blood pressure and physical fitness, smoking and alcohol consumption in Copenhagen males aged 40–59. *Acta Medica Scandinavica*, **195**, 375–80.

Jonsson, B.G. & Astrand, I. (1979). Physical work capacity in men and women aged 18–65. *Scandinavian Journal of Social Medicine*, **7**, 131–42.

Kleinbaum, D.G., Kupper, L.L. & Muller, K.E. (1988). *Applied Regression Analysis and Other Multivariable Methods*, Boston: PWS-KENT Publishing Co.

Lamb, K.L. & Brodie, D.A. (1991). Leisure-time physical activity as an estimate of physical fitness: a validation study. *Journal of Clinical Epidemiology*, **44**, 41–52.

Morris, J.N., Clayton, D.G., Everitt, M.G., Semmence, A.M. & Burgess, E.H. (1990). Exercise in leisure time: coronary attack and death rates. *British Heart Journal*, **63**, 325–34.

Peters, R.K., Cady, L.D., Bischoff, D.P., Bernstein, L. & Pike, M.C. (1983). Physical fitness and subsequent myocardial infarction in healthy workers. *Journal of the American Medical Association*, **249**, 3052–6.

Powell, K. (1987). Physical activity and the incidence of coronary heart disease. *Annual Review of Public Health*, **8**, 253–87.

Shephard, R.J. (1969). Learning, habitation, and training. *Internationale Zeitschrift für Angewandte Physiologie*, **28**, 38–48.

Shephard, R.J. (1970). Computer programs for solution of the Astrand nomogram and the calculation of body surface area. *Journal of Sports Medicine*, **10**, 206–10.

Shephard, R. (1974). Sudden death, a significant hazard of exercise. *British Journal of Sports Medicine*, **8**, 101–10.

Shephard, R. (1981). Present views on the Canadian Home Fitness Test. *Canadian Medical Association Journal*, **124**, 875–9.

Tuxworth, W., Nevill, A.M., White, C. & Jenkins, C. (1986). Health, fitness, physical activity, and morbidity of middle aged male factory workers I. *British Journal of Industrial Medicine*, **43**, 733–53.

Welsh Heart Programme Directorate (1987). Exercise for health: health-related fitness in Wales. *Heartbeat Report No. 23.*

Wenger, H.A. & Bell, G.J. (1986). The interactions of intensity, frequency and duration of exercise training in altering cardiorespiratory fitness. *Sports Medicine*, **3**, 346–56.

Wilhelmsen, L., Tibblin, G., Aurell, M., Bjure, J., Ekstrom-Jodal, B. & Grimby, G. (1976). Physical activity, physical fitness and risk of myocardial infarction. *Advances in Cardiology*, **18**, 217–30.

7 Physical development and childhood activity

PROF.DR. HAN C.G. KEMPER

Introduction

It has long been recognized that physical activity is an important consideration during the growing years if normal growth and development of children are to be maintained (Bar-Or, 1983). Children are generally thought to be naturally physically active. In recent years, however, the physical activity of youngsters has been a subject of great concern to health officials. Up to a generation ago, physical activity was a natural part of life for most children. This is no longer so, and one may well ask whether the child or the adolescent now gets the physical activity required for healthy development. The necessity for physical activity has been greatly reduced, owing to mechanization and automation of work and leisure. Currently, physical activity depends on such factors as body build, physical fitness, and the amount of recreational and sport facilities.

Physical inactivity is an important risk factor for coronary heart disease (CHD). Atherosclerosis starts soon after birth (Montoye, 1985). It is often suggested that a sufficient amount and intensity of regular physical activity could decelerate this process (Powell *et al.*, 1987). However, a prospective epidemiological study, comparing a large number of physically active children with a randomized group of less active children over a long period, has never been conducted and apparently cannot be carried out (Mednick & Baert, 1981). There is unfortunately no possibility of a double-blind study in which physical activity can be measured.

One way out of this dilemma is to measure habitual physical activity on a longitudinal basis, and to group individuals according to activity patterns. (Sprynarova, 1974; Rutenfranz *et al.*, 1974, 1975; Mirwald *et al.*, 1981). Another possibility is an experimental longitudinal study such as the Canadian one in the Trois Rivières region (Jéquier *et al.*, 1977) in which the effect of school classes with additional physical education was compared with control classes over a couple of school years.

In the course of body growth and development, there are critical periods that determine whether individuals will later lead a physically active life (Masironi & Denolin, 1985).

1. At age 4–12 years children first go to school and lose a considerable amount of playing time.
2. At age 12 years, children enter secondary school with further restriction of free time by homework.
3. At age 15–16 years, teenagers in industrialized countries shift from bicycles to motorcycles.
4. From age 18 years, young people further change to automobiles.

The consequences for the development of their daily energy expenditure is illustrated in Figure 7.1 by Rowland (1990).

The total daily energy expenditure (kcal/kg) diminishes very rapidly between 6 and 14 years of age. The decrease is almost 50% in both boys and girls. In the same figure, it can be seen that boys are more active than girls at all ages: the difference is about 10%. During these years lifestyle changes considerably and also their health perspective.

In this chapter the developmental changes in cardiorespiratory fitness of boys and girls during the teenage period are reviewed. The results of a Dutch longitudinal study in which maximal aerobic power and habitual physical activity were measured simultaneously are presented (Kemper, 1985; Kemper *et al.*, 1990; Kemper & Verschuur, 1990).

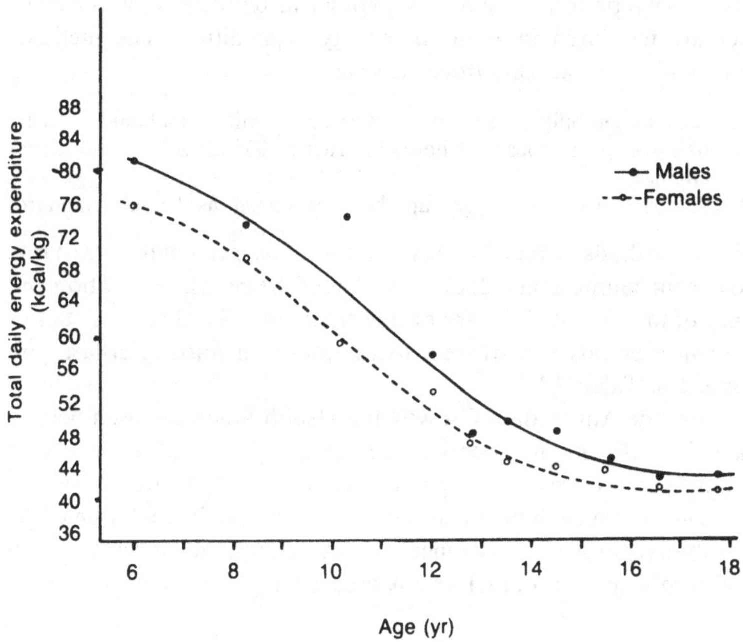

Figure 7.1. Total daily energy expenditure of boys and girls between the age of 6 and 18 years, © 1990 Human Kinetics Publishers, Inc. Reprinted with permission.

Table 7.1. *Characteristics of various methods of physical activity measurement*

	Criteria		
	Social acceptability	Applicability over 24 hr	Validity relative to $\dot{V}O_2$ measurement
Measurement methods			
Oxygen uptake	×	×	× × ×
Heart rate	× ×	× ×	× ×
Movement counter	× × ×	× × ×	×
Observational methods			
Diary	× ×	× ×	× ×
Questionnaire	× ×	× × ×	×

Rating scale: × × × = 'best'; × = 'worst'.

To evaluate the contribution of exercise to health and fitness of children, it is necessary not only to perform exercise tests but also to assess their daily physical activity pattern. From a physiological point of view, physical activities are measured in terms of energy expenditure. The methods adopted should fulfil at least three criteria:

1. social acceptability and minimal interference with normal daily activity;
2. applicable over at least 24 hours in order to include school and leisure activities;
3. valid with respect to oxygen uptake measurement as the gold standard.

Available methods include 1) measurements of oxygen uptake, heart rate and movement counters, and 2) observations of the intensity, duration and frequency of movements by diary or questionnaire. The characteristics of the different methods with respect to the above mentioned criteria are summarized in Table 7.1.

Data from the Amsterdam Growth and Health Study are used here to illustrate the different methods to measure daily physical activity in a longitudinal perspective of a group of teenage boys and girls from age 12 to 21 years. This chapter will be concluded by an analysis of the relation of the level of habitual physical activity measured as an independent variable with maximal aerobic power and CHD risk indicators as dependent variables.

Developmental aspects of maximal aerobic power during the teenage period

The commonly used index of maximal aerobic power is the highest volume of oxygen that can be consumed by the body per time unit. Maximal aerobic power ($\dot{V}O_2max$) has been extensively studied in both children and adults over the past 50 years. The early cross-sectional studies of Robinson (1938) and Åstrand (1952) provided the initial age-related benchmarks which later studies utilized to discuss and compare their results. Mirwald and Bailey (1985) argue that, despite an extensive bibliography, there are still gaps in the knowledge of age-related changes in $\dot{V}O_2max$. The majority of the investigations are cross-sectional, and longitudinal studies evaluating the same children from prepubertal to postpubertal stages have been few in number. Krahenbuhl, Skinner & Kohrt (1985) reviewed developmental aspects of maximal aerobic power in normal 6–16 year-old children by plotting the mean values reported in more than 60 studies from the literature. To correct for variation due to the mode of exercise, Krahenbuhl multiplied all values from cycle ergometer tests by a factor of 1.075. Using the regression line as a guide, it became clear that boys and girls differ in $\dot{V}O_2max$ from age 12 onwards, and that the differences grow larger during the teenage period. Most of their reviewed data, however, were from cross-sectional and not from longitudinal studies.

A survey is given of longitudinal studies that took into consideration the following: 1) serial measurements over a period of more than 3 years, 2) coverage of the adolescent period from 12 to 18 years of age, and 3) direct measurement of $\dot{V}O_2max$ (Table 7.2).

The Prague Growth Study (Sprynarova, 1974) was initiated in 1961 with 143 boys (average age 10.5 years). After 7 years, boys with complete data were regrouped according to overall time spent in physical activity. The Saskatchewan Growth and Development Study (Bailey, 1968) began in 1964 with a group of 7 year-old boys ($n = 131$), and, in 1965, 7 year-old girls ($n = 98$) were added, with the objective of following both groups for a period of 15 years. After 10 years of data collection for the boys and 9 years for the girls, financial support was withdrawn. Although preliminary results in boys have been published (Carron & Bailey, 1974; Mirwald, 1980), a considerable amount of data remained unanalysed. This study and the Prague study were the first to produce longitudinal data on maximal oxygen uptake of the same subjects over a considerable period of time. Cunningham *et al.* (1977, 1981) completed a 5-year longitudinal study on 81 boys, recruited from among participants in organized ice hockey (age 10–15 years), including annual tests of $\dot{V}O_2max$ on a treadmill. Inspired by the World Health Organization, during the 1970s Lange Andersen *et al.*

Table 7.2. *Longitudinal studies which have included repeated measurements of maximal aerobic power*

Name of the study	Sex	Age range (years)	Publication(s)
Prague Growth Study (CSSR)	Boys	11–18	Sprynarova (1974)
Saskatchewan Growth and Development Study (Canada)	Boys Girls	7–16 7–15	Bailey (1968) Carron & Bailey (1974) Mirwald (1980) Mirwald & Bailey (1985)
Canada Growth and Development Study (Canada)	Boys	10–15	Cunningham *et al.* (1977, 1981)
Two Countries Study (Norway, Germany)	Boys Girls	8–16 8–16	Lange Andersen *et al.* (1974) Rutenfranz *et al.* (1981)
Trois Rivières Regional Study (Canada)	Boys Girls	6–12	Jéquier *et al.* (1977) Shephard *et al.* (1980)
Amsterdam Growth and Health Study (The Netherlands)	Boys Girls	12–21	Kemper, 1985; Kemper & van't Hof, 1978; Kemper *et al.* (1983, 1990)

(1974) coordinated a longitudinal study of children in two countries, Norway and Germany: two small rural communities were chosen (Lom and Fredeburg). The children were followed from age 8 to 16 years. $\dot{V}O_2$max was measured (cycle ergometer) (Lange Andersen *et al.*, 1974; Rutenfranz *et al.*, 1981) as was its relationship to physical activity. The intention of the Canadian longitudinal study in the Trois Rivières region (Jéquier *et al.*, 1977) was not only to measure $\dot{V}O_2$max (treadmill) development of French Canadian schoolchildren but also to study the effect of additional physical education upon their development between the ages of 5 and 12 years (the normal one lesson per week was compared with five lessons per week). In the presence of many interfering factors such as experimental versus control groups, urban versus rural localities, it was not easy to evaluate the research hypotheses (Shephard *et al.*, 1980). Kemper and colleagues conducted a multiple longitudinal study of Dutch teenagers from 12/13 and 17/18 years of age. The design consisted of four birth cohorts, measured during 4 years. $\dot{V}O_2$max of 102 boys and 131 girls was measured directly using a standard treadmill protocol (Kemper & van't Hof, 1978; Kemper *et al.*, 1983).

Figure 7.2 illustrates the developmental trends of $\dot{V}O_2$max in boys and girls in the teenage period. It comprises studies from Canada (Mirwald & Bailey, 1984), Germany (Rutenfranz *et al.*, 1981), the Netherlands (Kemper, 1985) and Norway (Lange Andersen *et al.*, 1974). The data are related to body weight as a functional parameter of maximal aerobic power ($\dot{V}O_2$max/BW). At age 12, the mean value for boys is 56 ml/min/kg. This decreases markedly in German boys, less in Norwegian boys and not at all in Dutch boys. The $\dot{V}O_2$max/BW of girls is on average 50 ml/min/kg at age 12 and decreases in all the studies till the end of the teenage period. There is a consistent trend of a greater decrease of $\dot{V}O_2$max in girls than in boys

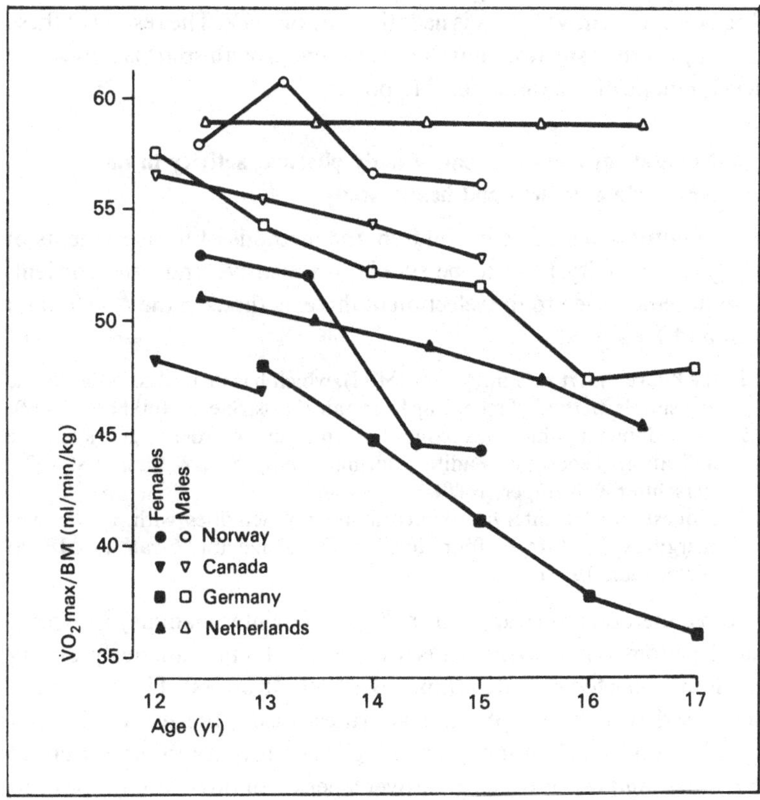

Figure 7.2. Longitudinal data from four studies of maximal aerobic power relative to bodyweight in boys and girls in the teenage period.

within each study. The decrease in adolescent girls stems predominantly from the increase in body fat.

Longitudinal measurement of $\dot{V}O_2max$ allow the study of individual changes in growth and development. In earlier research it has been demonstrated repeatedly that there is a strong interrelationship between the time of peak height velocity (PHV) and that of other pubertal changes in boys and girls. In representing developmental curves of some characteristics, a scale before and after PHV can also be used. In the relevant literature, this procedure was used for anthropometric data by Tanner, Hughes & Whitehouse (1981) and Marshall (1977) and for data on strength by Carron and Bailey (1974) and Beunen & Malina (1988). Kobayashi *et al.* (1978) were the first to present $\dot{V}O_2max$ data versus age at PHV in a longitudinal training study. Mirwald *et al.* (1981) and Andersen & Froberg (1980) also used the PHV as a point of reference in their longitudinal training experiments with boys. In the study of Kemper (1985) the girls' greatest increase in $\dot{V}O_2max$ generally occurs one year before PHV. The boys' peak increase in $\dot{V}O_2max$ is near the time of PHV. The results of these studies support the assertion that the adolescent growth spurt is critical for the development of maximal aerobic power.

Methods of measurement of daily physical activity in the Amsterdam growth and health study

In large longitudinal studies in children, the methods of measurements of daily physical activity have to be simple, inexpensive and time efficient. These requirements led to the selection of three methods in the Amsterdam Growth and Health Study:

1. the 8-level heart rate integrator (HRI), which has proved to be a reliable and simple method of recording heart rate (Saris, Snel & Binkhorst, 1977).
2. the pedometer, which was reduced in sensitivity in order to give a reliable indication of energy expenditure during running (Saris & Binkhorst, 1977; Verschuur & Kemper, 1980).
3. a questionnaire–interview aimed at tracing activities with a minimum energy expenditure of four times the basal metabolic rate (4 METS) (Verschuur, 1987).

The three methods were applied each year in winter (January to April). HRI and pedometer measurements were applied simultaneously during two randomly selected weekdays (approximately 48 hours). The pedometer was also used to measure physical activities during the weekend (from Friday afternoon until Monday morning). The activity interview lasted 10–15 minutes and was retrospective over a period of three months prior to the interview.

The HRI is an electronic system that measures, analyses and stores R-R

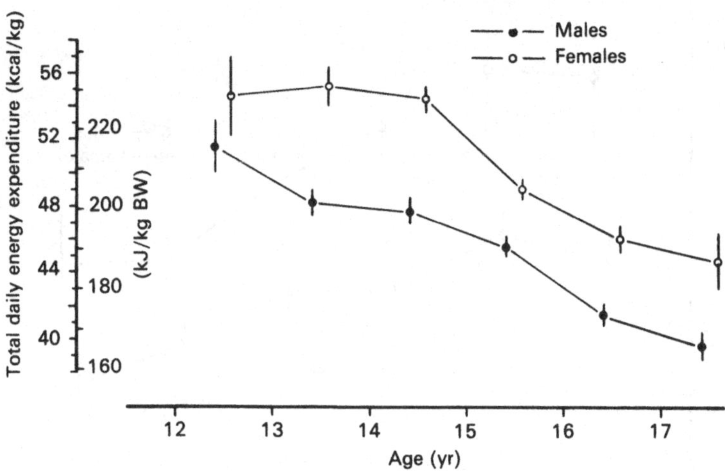

Figure 7.3. Mean and standard error of daily energy expenditure relative to bodyweight in Dutch boys and girls calculated from 48-hour heart rate measurements.

intervals of the ECG in one of the eight registers corresponding to different 25–30 beat heart rate (HR) ranges. To transfer heart rate into aerobic energy expenditure, the linear relationship between HR and $\dot{V}O_2max$ of each subject measured from a submaximal treadmill test was used. Since energy consumption is determined largely by body weight, daily energy expenditure in boys and girls during teenage period is given as kcal or kJ relative to body weight in Figure 7.3.

In both girls and boys daily energy expenditure decreases from 12–17 years of age. Boys have significantly higher energy expenditure than girls at pre-pubertal ages (13 and 14 years) but post-pubertally the differences become very small and are not significantly different.

The pedometer week score is shown in Figure 7.4.

In girls and boys there is a decline in pedometer score of 52% and 44% respectively from age 12 to age 17 years. At any age, the week score of the boys is significantly higher than the girls. To classify activities according to their energy expenditure independent of body weight, the ratio between work metabolic rate and basal metabolic rate, the so called metabolic equivalent (MET) was used. All activities scored were sub-divided into three levels of intensity: light (4–7 METS), for example, carrying loads, bicycling and dancing; medium heavy (7–10 METS), for example, stair-climbing, swimming, tennis; and heavy activities (>10 METS), for example, running, playing soccer and rowing. Through the interview, the average weekly time spent in each intensity level was assessed and a weighted activity score, combining duration and intensity calculated. The

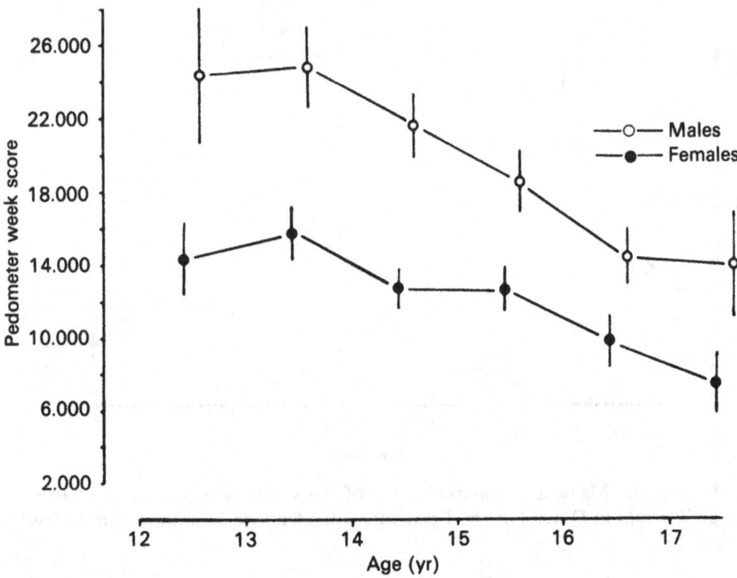

Figure 7.4. Mean and standard error of the pedometer week score, calculated from 2 weekdays + 1 weekend measurement period in Dutch boys and girls.

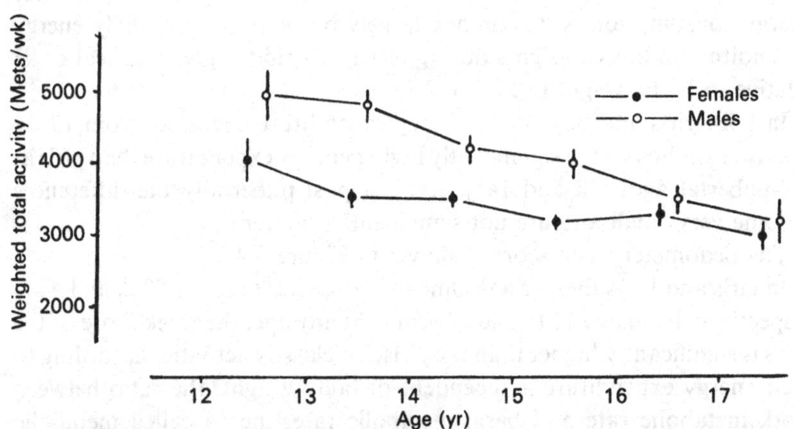

Figure 7.5. Mean and standard error of the weighted activity score, the total number of METs per week spent above a level of 4 METs, in Dutch girls and boys.

Table 7.3. *Pearson correlations between the three activity instruments: mean values of four years of measurement. In the upper right triangle values of boys, in the lower left triangle values of girls*

	HRI	Pedometer	Interview
Total energy weekscore from heart rate indicator (HRI)	×	0.17	0.16
Total weekscore from pedometer	0.16	×	0.20
Total weighted activity from interview	0.18	0.20	×

time spent per level of intensity was multiplied by a fixed value of each intensity level: light (5.5), medium (8.5) and heavy (11.5). The total scores of the three levels were added to a total METs per week score. The longitudinal changes with age and sex of this weighted activity score are illustrated in Figure 7.5.

In both girls and boys the weighted activity scores decrease gradually during the teenage period.

The information from the three activity methods overlaps only partly. The HRI was applied for 48 hours only, the pedometer during the same days plus one weekend and the activity interview looked back over a period of the previous three months. Furthermore, there is a difference in the intensity and kind of activities that can be evaluated by the three methods: pedometers measure only vertical changes of the whole body such as in running, but not in walking or during bicycling.

The correlation coefficients between the activity variables from the three approaches are given in Table 7.3.

The Pearson correlation coefficients in boys and girls, averaged over 4 years of measurement, vary between 0.16 and 0.20 explaining only a very small part of total variation. This illustrates that it is not possible to use only one of the three methods, and that a combination of difference measurements and observational methods is needed to obtain a valid picture of children's daily activity pattern.

Relationship of daily physical activity with maximal aerobic power and CHD risk indicators

To investigate the relationship between the level of habitual physical activity as an independent variable with $\dot{V}O_2$max as the dependent variable, all girls and boys measured during the first 4 years into the

Amsterdam Growth and Health Study were divided into 'active' and 'inactive' groups (Verschuur, 1987). The selection was on the basis of three activity variables: 1) energy expenditure measured with heart rate recording over 48 h, 2) pedometer week scores, and 3) total time spent per week according to an activity interview. The active individuals were those who scored above the median in at least 3 out of 4 years. The inactive were those who scored below the median in 2 out of 4 years. The difference between these activity groups was primarily in the high-intensity activities. Comparing the active and inactive pupils, there were no differences in height, weight and fat-free mass. In contrast, parameters directly related to cardiorespiratory fitness were significantly affected by the level of physical activity: the active girls and boys have higher $\dot{V}O_2max/BW$ than the inactive ones. Again, the differences are small and even the inactive ones have a reasonably high aerobic fitness (Figure 7.6). Since the differences in $\dot{V}O_2max/BW$ between the active and the inactive youngsters already exist at age 12/13 years self selection rather than training could be the cause of these differences as indicated by Rutenfranz and Singer (1980).

Although it is generally assumed that, in western industrialized countries, aerobic power has declined during the past 25 years, data do not support this assumption: the mean values of the boys and girls in the Netherlands are similar to those found in studies from 1938 (Robinson) and 1952 (Åstrand). The results of the Amsterdam Growth and Health Study agree with those of Kobayashi *et al.* (1978) and Mirwald *et al.* (1981): physical activity or training has no significant effect on $\dot{V}O_2max$ prior to adolescence. In these studies the active subject's $\dot{V}O_2max$ is higher than the inactive subject's, but only after the pubertal growth spurt do the differences attain statistical significance (Mirwald & Bailey, 1984). The

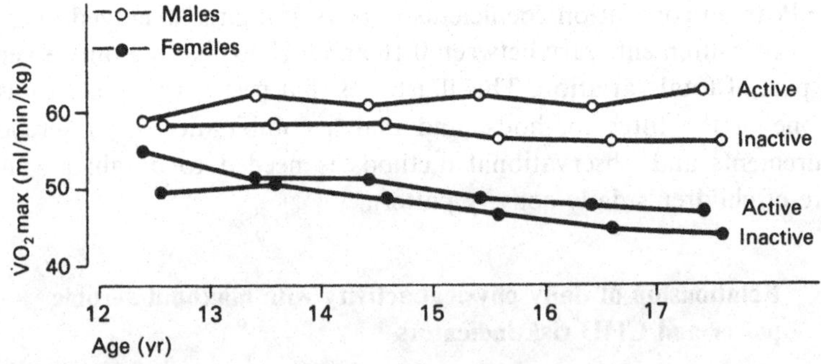

Figure 7.6. Mean values of maximal aerobic power relative to body weight (ml/kg min) in active and inactive Dutch boys and girls divided longitudinally on the basis of daily activity.

authors present the following reasons for the greater increase after PHV: superior genetic endowment, habitual physical activity pattern, increased intensity of physical training and testosterone secretion during adolescence.

So far, the results of the Amsterdam Growth and Health Study are from a population of 93 males and 107 females who were measured annually from 1977 to 1980. However, in 1985, a fifth measurement was made in the same population at age 21. In that way, longitudinal data covering a period of 8 years were collected for a group of adolescents/adults between 13 and 21 years of age. Both genetic and behavioural lifestyle parameters were measured as indicators of CHD: serum cholesterol, blood pressure, percentage body fat (from four skinfolds), maximal aerobic power, smoking, type A/B behaviour and daily physical activity (Kemper *et al.*, 1990). Physical activity was measured by the same interview as in the foregoing years. For each of the seven CHD indicators, groups were formed of males and females who were either above the fiftieth percentile (P50) every year of the study or below it on each occasion. The group means of the three environmental CHD factors, smoking, type A/B behaviour and daily physical activity at age 21 years in the same sex upper and lower groups, were tested for significance of differences. The results are summarized in Table 7.4. Physical activity was the only parameter that showed significant differences ($P < 0.05$) in several of the groups based on the seven other CHD indicators. Higher levels of activity were found in boys and girls in the groups with higher HDL-C, higher $\dot{V}O_2max$ and lower percentage body fat. These results indicate the importance of daily physical activity in adolescence in reducing CHD risk factors in young adulthood.

Bell *et al.* (1986) suggested that the development of a health-related profile would facilitate the identification of individuals who, at an early age, are already developing risk factors that might lead to complications in adult life. In Table 7.5 the health and risk criteria are summarized.

To establish the stability of belonging to a relatively high- or low-risk group, each year the sample was split at the median (P50). Subjects who were in the upper half (>P50) each of the 5 years of measurement were referred to as being at relatively high risk for CHD; those appearing in the lower half (<P50) each year of measurement were seen as being at relatively low risk for CHD. The relatively high risk groups show, in general, mean values that are well below the risk values of Bell *et al.* (1986) and more in the direction of health values with one exception. Boys and girls in the relatively high groups of percentage of body fat (Figure 7.7) have already during the teenage period mean values that are equal to, or even higher than, the risk values of 30% fat in girls and of 20% in boys.

At 21 years of age, relatively high risk males and females have a

Table 7.4. *Statistical significance (t-test, one-sided) of the difference in mean values for smoking, physical activity and type A/B behaviour in groups of males and females selected on the basis of being above or below the median throughout the study for the seven CHD indicators (selection criteria)*

Tested CHD indicators	Selection criterion						
	TC	HDL	TC/HDL	% fat	P_{syst}	P_{diast}	$\dot{V}O_2max$
Smoking							
males			$P<0.05$				
females							
Physical activity							
males		$P<0.01$	$P<0.01$	$P<0.01$			$P<0.01$
females		$P<0.05$		$P<0.01$			$P<0.01$
Type A/B							
males							
females				$P<0.05$			

TC = Total cholesterol
HDL = High density lipoprotein cholesterol
TC/HDL = Ratio between TC and HDL
P_{syst} = Systolic blood pressure
P_{diast} = Diastolic blood pressure

Table 7.5. *Adolescent health and risk levels from the European Society of Pediatric Work Physiology, proposed by Bell et al. (1986)*

	Health level	Risk level
Diastolic blood pressure (P_{diast})	75–85 mm Hg	>90 mm Hg
Systolic blood pressure (P_{syst})	100–20 mm Hg	>140 mm Hg
Total serum cholesterol (TC)	≤4.9 mmol/l	>5.9 mmol/l
High density serum lipoprotein (HDL)	≥1.3 mmol/l	≤0.8 mmol/l
TC/HDL ratio	≤2.7	≥5.5
Percentage of body fat		
males	<15%	>20%
females	<20%	>30%
Maximal aerobic power ($\dot{V}O_2max$/BW)		
males	>40 ml/min kg	<35 ml/min kg
females	>35 ml/min kg	<30 ml/min kg

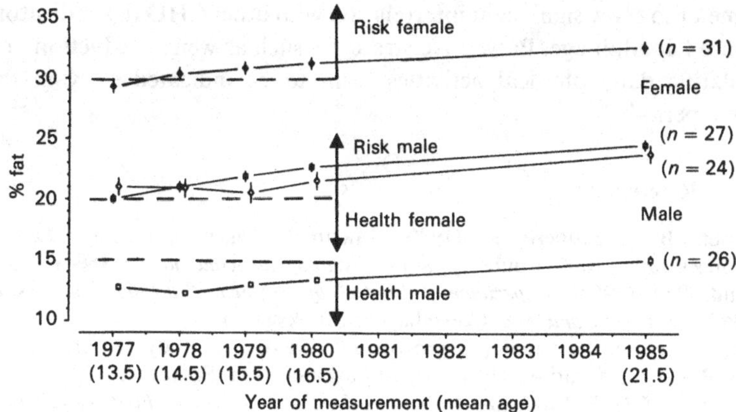

Figure 7.7. Mean and standard error of percentage of body weight as fat in relatively low (<P50) and relatively high (>P50) risk groups for CHD in Dutch boys and girls measured from age 13 to 21 years. Relatively high risk (above the median on five occasions) boys: open diamond; girls: closed diamond; relatively low risk (below median on five occasions) boys: open square; girls: closed square. These adolescent risk and health values indicated are according to Bell *et al.* (1986), as shown in Table 7.5. ↓: less than; ↑: more than.

percentage of body fat of 24% and 33% respectively. Even the relatively low risk males and females have, at this age, mean values that are outside the healthy range.

Conclusions

The physical development of maximal aerobic power during the teenage period in six longitudinal studies shows a consistent general trend of a greater decrease in $\dot{V}O_2$max relative to body weight in girls than in boys. At the same time, these data do not support a decline in aerobic power during the past 25 years.

Longitudinal measurements of daily physical activity show that boys are more active than girls, especially in heavy activities (>10 METs). In both sexes the amount of activity decreases from 12 to 18 years dramatically.

Relatively active children show higher $\dot{V}O_2$max values than their relatively inactive counterparts. The results, however, give more support to the hypothesis that children during their teens are active because they have high aerobic power than the reverse, that children have high aerobic power because they are active.

The percentage body fat in the early teenage period seems to be the most important coronary heart disease indicator in predicting risk levels in the young adult. The amount of daily physical activity is the only behavioural parameter to show significant interrelation with other CHD risk indicators measured at adult age. Preventive strategies such as weight reduction and stimulating daily physical activities seem to be indicated early in the teenage period.

References

Andersen, B. & Froberg, K. (1980). Maximal oxygen uptake and lactate concentration during puberty. *Acta Physiologica Scandinavica*, **188**, 6.

Åstrand, P.O. (1952). *Experimental Studies of Physical Working Capacity in Relation to Sex and Age*. Copenhagen: Munksgaard.

Bailey, D.A. (1968). Saskatchewan growth and development study. Report College of Physical Education, University of Saskatchewan, Saskatoon.

Bar-Or, O. (1983). *Pediatric Sport Medicine for the Practitioner: From Physiologic Principles to Clinical Applications*. New York: Springer.

Bell, R.D., Macek, M., Rutenfranz, J. & Saris, W.H.M. (1986). Health factors and risk indicators of cardiovascular diseases during childhood and adolescence. In *Children and Exercise XII*. Rutenfranz, J., Mocellin, R. and Klimt, F., eds, pp. 19–27. Champaign: Human Kinetics.

Beunen, G. & Malina, R.M. (1988). Growth and physical performance relative to the timing of the adolescent spurt. *Exercise and Sport Sciences Reviews*, **16**, 503–40.

Carron, A.V. & Bailey, D.A. (1974). Strength development in boys from 10 through 16 years. *Monographs of the Society for Research in Child Development*, **157**, 39.

Cunningham, D.A., Stapleton, J.J., MacDonald, I.C. & Paterson, D.H. (1981). Daily expenditure of young boys as related to maximal aerobic power. *Canadian Journal of Applied Sports Science*, **6**, 207–11.

Cunningham, D.A., Waterschoot, B.M. van, Patterson, D.H., Lefcoe, M. & Sangal, S.P. (1977). Reliability and reproducibility of maximal oxygen uptake measurements in children. *Medicine and Sport Science*, **9**, 104–8.

Jéquier, J.C., Lavallee, H., Rajic, M., Beaucage, L., Shephard, R.J. & Labarre, R. (1977). The longitudinal examination of growth and development: history and protocol of the Trois Rivères regional study. In *Frontiers of Activity and Child Health*. Lavallee, J. and Shephard, R.J., eds, pp. 49–54. Ottawa: Pelican.

Kemper, H.C.G. (1985). *Growth, Health and Fitness of Teenagers. Longitudinal Research in International Perspective*. Basel: Karger.

Kemper, H.C.G., Dekker, H.J.P., Ootjers, M.G., Post, G.B., Snel, J., Splinter, P.G., Storm-van Essen, L. & Verschuur, R. (1983). Growth and health of teenagers in the Netherlands. Survey of multidisciplinary longitudinal studies and comparison to recent results of a Dutch study. *International Journal of Sports Medicine*, **4**, 202–14.

Kemper, H.C.G. & Hof, M.A. van't (1978). Design of a multiple longitudinal study of growth and health in teenagers. *European Journal of Pediatrics*, **129**, 147–55.

Kemper, H.C.G., Snel, J., Verschuur, R. & Storm-van Essen, L. (1990). Tracking of health and risk indicators of cardiovascular diseases from teenager to adult: Amsterdam Growth and Health Study. *Preventive Medicine*, **19**, 642–55.

Kemper, H.C.G. & Verschuur, R. (1990). Longitudinal study of coronary risk factors during adolescence and young adulthood: the Amsterdam Growth and Health Study. *Pediatric Exercise Science*, **12**, 359–71.

Kobayashi, K., Kitamure, K., Miura, M., Sodeyama, H., Murase, Y., Miyashita, M. & Matsui, H. (1978). Aerobic power as related to body growth and training in Japanese boys: a longitudinal study. *Journal of Applied Physiology* **44**, 666–72.

Krahenbuhl, G.S., Skinner, J.S. & Kohrt, W.M. (1985). Developmental aspects of maximal aerobic power in children. *Exercise and Sport Sciences Reviews*, **13**, 503–38.

Lange Andersen, K., Seliger, V., Rutenfranz, J. & Mocellin, R. (1974). Physical performance capacity of children in Norway. Part 1. Population parameters in a rural inland community with regard to maximal aerobic power. *European Journal of Applied Physiology*, **33**, 177–95.

Marshall, W.A. (1977). *Human Growth and its Disorders*. London: Academic Press.

Masironi, R. & Denolin, H. (1985). *Physical Activity in Disease, Prevention and Treatment*. Padua: Piccin.

Mednick, J.A. & Baert, A.E. (1981). *Prospective Longitudinal Research: An Empirical Basis for the Primary Prevention of Psychological Disorders*. Oxford: Oxford University Press.

Mirwald, R.L. (1980). Saskatchewan growth and developmental study. In *Kinanthropometry II*. Ostyn, M, Beunen, G. and Simons, J., eds, pp. 289–305. Baltimore: University Park Press.

Mirwald, R.L. & Bailey, D.A. (1984). Longitudinal comparison of aerobic power and heartrate responses at submaximal and maximal workloads in active and inactive boys aged 8 to 16 years. In *Human Growth and Development*. Borms, J., Hauspie, J., Sand, A., Susanne, C. and Hebbelinck, M., eds, pp. 561–69. New York: Plenum.

Mirwald, R.L. & Bailey, D.A. (1985). *Longitudinal Analysis of Maximal Aerobic Power in Boys and Girls by Chronological Age, Maturity and Physical Activity*. Saskatoon: University of Saskatchewan.

Mirwald, R.L., Bailey, D.A., Cameron, N. & Rasmussen, R.L. (1981). Longitudinal comparison of aerobic power in active and inactive boys aged 7.0 to 17.0 years. *Annals of Human Biology*, **8**, 405–14.

Montoye, H.J. (1985). Risk indicators for cardiovascular disease in relation to physical activity in youth. In *Children and Exercise XI*. Binkhorst, R.A., Kemper H.C.G. & Saris W.H.M., eds., pp. 3–25. Champaign: Human Kinetics.

Powell, K.E., Thomson, P.D., Caspersen, C.J. & Kendrick, J.S. (1987). Physical activity and the incidence of coronary heart disease. *Annual Review of Public Health*, **8**, 253–87.

Robinson, S. (1938). Experimental studies of physical fitness in relation to age. *Arbeitsphysiologie*, **10**, 252–323.

Rowland, T.W. (1990). *Exereise and Children's Health*. p. 35. Champaign: Human Kinetics.

Rutenfranz, J., Andersen, K.L., Seliger, V., Berndt, I. & Rupper, M. (1981). Maximal aerobic power and body composition during the puberty growth period: similarities and differences between children of two European countries. *European Journal of Pediatrics*, **13**, 123–33.

Rutenfranz, J., Berndt, I. & Knauth, P. (1974). Daily physical activity investigated by time budget studies and physical performance capacity of schoolboys. In *Children and Exercise*, Borms, J. and Hebbelinck, M., eds, pp. 79–86. *Acta Paediatrica Belgica*, **28**, suppl.

Rutenfranz, J., Seliger, V., Andersen, K.L., Ilmarinen, J., Berndt, I. & Knauth, P.

(1975). Differences in maximal aerobic power related to the daily physical activity in childhood. In *Physical Work and Effort*, Borg, G.V., ed., pp. 279–88. Oxford: Pergamon.

Rutenfranz, J. & Singer, R. (1980). The influence of sport activity on the development of physical performance capacities of 15–17 year old boys. In *Children and Exercise IX*, Berg, B. and Erikson, B., eds., pp. 160–65. Baltimore: University Park Press.

Saris, W.H.M. & Binkhorst, R.H. (1977). The use of pedometer and actometer in studying daily physical activity in man. *European Journal of Applied Physiology*, **37**, 219–35.

Saris, W.H.M., Snel, P. & Binkhorst, P.A. (1977). A portable heart rate distribution recorder for studying daily physical activity. *European Journal of Applied Physiology*, **37**, 19–25.

Shephard, R.J., Lavallee, H., Jéquier, J.C., Rajic, M. & Labarre, R. (1980). Additional physical education in the primary school – a preliminary analysis of the Trois Rivières regional experiment. In *Kinanthropometry II*. Ostyn, M., Beunen, G. and Simons, J., eds, pp. 306–16. Baltimore: University Park Press.

Sprynarova, S. (1974). Longitudinal study of the influence of different physical activity programs on functional capacity of the boys from 11–18 years. *Acta Paediatrica Belgica*, **29**, suppl., 204–13.

Tanner, J.M., Hughes, P.C.G. & Whitehouse, R.H. (1981). Radiographically determined width of bone, muscle and rat in the upper arm and calf from age 3–18 years. *Annals of Human Biology*, **8**, 495–517.

Verschuur, R. (1987). Habitual physical activity and health. In *Longitudinal Changes During the Teenage Period.* Haarlem: De Vrieseborch.

Verschuur, R. & Kemper, H.C.G. (1980). Adjustment of pedometers to make them more valid in assessing running. *International Journal of Sports Medicine*, **1**, 87–9.

8 *Physical activity and behavioural development during childhood and youth*

ROBERT M. MALINA

The developmental tasks of children and youth are many. Their successful attainment is related in part to the child's growth, maturation and development, and to numerous factors in the child's environments. One of the major tasks of childhood and youth is the development of competence in a variety of behavioural domains. Although many factors are involved in the development of behavioural competence, physical activity is often viewed as having a favourable influence. The role of physical activity in the development of several aspects of behavioural competence during childhood and youth is the focus of this chapter.

Growth, maturation and development

The terms growth, maturation and development often have different meanings to professionals involved with children and youth. Suggested domains of each are outlined in Table 8.1. Growth refers to changes in size, physique and body composition, and in various systems of the body, many of which are related to growth in body size, e.g. the cardiovascular and respiratory systems.

Maturation refers to the rate of progress towards the mature state. When dealing with behaviour, maturation is most often viewed in the context of sexual and somatic maturation during adolescence. Individuals attain skeletal, sexual and somatic maturity, but do so at varying rates. The neuromuscular domain is an additional indicator of maturity, particularly in infancy and early childhood. Early motor development and level of proficiency in fundamental movement tasks may serve as an indicator of neuromuscular maturation.

Development refers to the acquisition of competence in a variety of behavioural domains within the traditions and sanctions set by the particular culture within which the individual is reared. Children develop competence in several interrelated domains as they progress through

Table 8.1. *Domains of growth, maturation, and development*

Growth	Maturation	Development
Size	Skeletal	Cognitive
Physique	Sexual	Emotional
Composition	Somatic	Social
Systemic →	Neuromuscular	← Motor
	↓	
	Self-concept	
	↓	
	Perceived competence	

childhood and adolescence, i.e. social competence, cognitive competence, emotional competence, motor competence, and perhaps others. Note that the motor domain is included in both maturation and development. Motor development during infancy and early childhood leading to the establishment of fundamental movement patterns is conditioned largely by the interaction of the genotypically determined rate of neuromuscular maturation and multiple environments, e.g. rearing, nutrition, sibling interactions, and so on.

Growth and maturation are biological processes, while development is primarily a behavioural process. The three processes are not independent; rather, they are interdependent, which emphasizes the need to approach children in a biobehavioural or biocultural context. This is especially apparent, for example, in the maturity-associated variation in body size, strength, motor performance, and behaviour that is characteristic of male adolescence.

A child's growth, maturity and developmental characteristics interact to mould his/her self-concept (perception of self) and self-esteem (value placed on one's self-concept). Physical activity and self-esteem are discussed in more detail below.

Competence

Competence is the capacity to effectively interact with, and become adapted to, the cultural milieu of which the individual is a member. It implies the capacity to function effectively within a specific cultural pattern. Competence can be defined in a variety of ways, depending upon one's perspective(s).

Physical activity

Physical activity refers to '. . . any bodily movement produced by skeletal muscles and resulting in energy expenditure' (Bouchard *et al.*, 1990, p. 6). It clearly has many forms and contexts: free movement, play, exercise, dance, physical education, sport, and others. The ability to perform physical activities requires some degree of motor skill and physical fitness. Skill emphasizes the accuracy, precision and economy of performance, while fitness refers to the ability to perform muscular work over a period of time without diminished efficiency. Regular participation in physical activities can lead to an improvement in skill and fitness. Thus, any discussion of physical activity as a factor influencing behavioural competence in children and youth must consider motor skills and physical fitness.

Motor competence

The development of motor competence is a primary task of infancy and childhood. All children, except some with developmental complications, have the potential to develop and learn a variety of fundamental movement patterns and more specialized skills. Such motor activities are an integral part of children's behavioural repertoires, and provide the medium through which children experience many dimensions of their environments.

The development of fundamental movement patterns (in addition to walking, running, jumping, throwing, kicking, hopping and so on) is a process of continuous modification based upon the interaction of the child's genotypically mediated rate of neuromuscular maturation, residual effects of prior movement experiences, and the new motor experiences *per se*. Movement experiences are influenced by rearing conditions, which vary among cultures. The child's physical growth characteristics also influence the developmental process. Between 2 and 6 years of age when many basic movement patterns are emerging, rate of growth in stature and weight (though declining) is quite rapid, body proportions are changing, and subcutaneous fatness is decreasing (see Malina & Bouchard, 1991).

Learning and practice are significant factors affecting motor competence. Although the benefits of instructional programmes for preschool children have been questioned, such programmes can, in fact, enhance the development of fundamental motor skills. Essential components of successful programmes at young ages include guided instruction by specialists and/or trained parents, appropriate motor task sequences, and adequate time for practice (Haubenstricker & Seefeldt, 1986).

By 6 to 8 years of age, most children have developed basic levels of proficiency in fundamental movement patterns, although mature patterns of some skills do not develop until later (Seefeldt & Haubenstricker, 1982). It is also at these ages that motor competence becomes performance oriented as fundamental movement patterns are modified and/or combined into specific skills that are used in games, sports, dance, and other physical activities. Instruction and practice have an important role in the transition from fundamental movement patterns to specific skills.

Physical fitness

Although physical fitness presently has a two-fold context: motor fitness and health-related fitness, emphasis on the fitness of children and youth has shifted from a primary motor focus to a health-related focus over the past 15 years or so (Malina, 1991). Motor fitness is performance oriented and includes components of skilled movements, i.e. agility, balance, coordination, power, speed, strength and muscular endurance that enable the individual to perform a variety of physical activities (American Alliance for Health, Physical Education and Recreation, 1976; Canadian Association for Health, Physical Education and Recreation, 1980; Simons & Renson, 1982). It is a major component of motor competence.

Health-related fitness is operationalized in terms of cardiorespiratory function, abdominal and low back musculoskeletal function, and body composition, specifically subcutaneous fatness (American Alliance for Health, Physical Education, Recreation and Dance, 1980, 1984). Being aerobically fit, strong, lean and flexible is viewed as healthy and presumably reduces the risk of cardiovascular and perhaps other degenerative diseases. The shift in emphasis to health-related fitness for children and youth is based on either one or both of the following premises: 1) regular physical activity during childhood and youth may function to prevent or impede the development of several adult diseases that include physical inactivity in a complex, multifactorial etiology: degenerative diseases of the heart and blood vessels, obesity, musculoskeletal disorders (specifically low back pain), and perhaps others; and 2) habits of regular physical activity during childhood and youth may directly and favourably influence physical activity habits in adulthood and, in turn, have a beneficial influence on the health-related fitness and thus the health status of adults (Malina, 1991).

Motor and health-related fitness obviously have a major behavioural component, i.e. physically active behaviour. Regular participation in physical activity is essential to the development and maintenance of fitness. In children and youth, however, indicators of motor and health-related fitness change as a function of growth and maturation, and several show distinct growth spurts particularly in males (Beunen *et al.*, 1988: Beunen &

Table 8.2. *Reasons given by children and youth to participate in organized sports*[a]

	Boys (%)	Girls (%)
To have fun	93	96
To learn new skills	80	81
To become physically fit	56	56
To be with my friends	55	50
To make new friends	33	44

[a]Adapted from Sapp and Haubenstricker (1978).

Malina, 1988; Malina & Bouchard, 1991). Hence, it may be difficult to partition the influence of physical activity from that of normal growth and maturation. Nevertheless, physically active youth tend to demonstrate higher levels of motor and health-related fitness than inactive youth (Ruffer, 1965; Parizkova, 1977; Watson & O'Donovan, 1977; Schmucker *et al.*, 1984; Mirwald & Bailey, 1986; Pate & Ross, 1987; Moravec, 1989; Pate, Dowda & Ross, 1990).

Determinants of physical activity

There is a need for systematic study of factors which influence or determine interest in, motivation for, and habits of physical activity during childhood and youth. Daily physical education is not a feature of the lifestyle of the majority of school children. Physical activity outside of the school setting thus plays an increasingly important role.

It is not surprising, therefore, that organized sport is a major form of physical activity in school children. A small percentage of children begin participating in organized sport at 5 years of age, but by 7 or 8 years of age, over 50% have already begun to participate (Seefeldt & Haubenstricker, 1988). The two primary reasons stated by most children for participation in sports are to have fun and to learn new skills (Table 8.2). Social concerns, i.e. to be with friends or to make new friends, and physical fitness are also important reasons of many children.

Participation in organized youth sports reaches a peak at about 12–13 years of age, and then declines. The major reasons for the decline are the desire to try new things and changing interests (Sapp & Haubenstricker, 1978; Gould & Horn, 1984), although more rigid selection and increased specialization associated with sport, especially school sport, may be contributory factors. These trends for the United States find a parallel in observations on Finnish youth, which suggest a change in motivation for physical activity from emphasis on performance- and competition-oriented motivation in early adolescence (11–13 years) to recreational motivation in late adolescence (Telama & Silvennoinen, 1979). Among a sample of primarily first-year college students of both sexes, the three most commonly stated reasons why involvement in physical activity was very important are to feel better mentally and physically, to control weight and look better, and for pleasure, fun and excitement (Wellens, 1989).

The trend in youth sports participation also has a parallel in estimated levels of physical activity in national surveys in the United States and Canada and in several longitudinal studies in Europe. Activity levels increase, on average, from middle childhood into mid-adolescence in both sexes; subsequently, they tend to decline (Malina, 1986). Data for American and Dutch children and youth are presented in Tables 8.3 and 8.4, respectively. The adolescent decline in physical activity occurs after the growth spurt and is probably related to the social demands of adolescence, changing interests, and perhaps the transition from school to work or college.

The decline in physical activity during adolescence is apparently not a recent phenomenon as often implied in general comments on reduced levels of habitual activity and fitness in contemporary youth. Results of a survey from the early 1930s are summarized in Table 8.5. Weekly time spent as a participant in physical play decreased by about 36% between 12 and 16 years of age in boys, while weekly time spent as a spectator of physical play more than doubled (Dimock, 1935).

In the world of children and youth, physical activities have different meanings and significance that are situation specific. Successful performance of a basic skill by a 5 or 6 year-old child, or the ability to perform an invigorating dance by a 12 or 13 year-old boy has meaning far beyond its physiological or neuromuscular effects. Further, physical activities are commonly used in teaching a variety of subjects, e.g. manipulative activities (gross and fine motor) in mathematics in the primary grades, folk dance in social or multicultural studies, and so on. And, these aspects of physical activity are often overlooked.

There are thus multiple forms of physical activity. Their effects may be additive physiologically in terms of energy expenditure, but are probably quite differential behaviourally. The influence of physical activities on behavioural development depends on the context of the activities.

Table 8.3. *Weekly time (hours) in physical activity outside of school physical education in the US National Children and Youth Fitness Study, by grade and sex*[a]

Grades	Approximate age	Males	Females
5–6	10–11	13.7	11.5
7–9	12–14	13.8	12.5
10–12	15–17	12.5	11.8

[a]Adapted from Ross *et al.* (1985)

Table 8.4. *Average total activity time (hours per week) in physical activity with an energy expenditure of 4 METs*[a] *or more, including school physical education, in Dutch youth, mixed-longitudinal, 12 to 18 years of age*[b]

Age	Males	Females
12–13	10.4	9.6
13–14	10.2	8.5
14–15	9.2	8.6
15–16	9.5	8.3
16–17	8.3	8.8
17–18	7.7	8.1

[a]METS = multiples of basal metabolism.
[b]Adapted from Kemper *et al.* (1983).

Table 8.5. *Average hours per week in physical play as a participant and as a spectator by boys 12 to 16 years of age*[a]

Age	*n*	Participant	Spectator
12	30	10.5	0.7
13	61	9.6	1.2
14	83	7.7	1.4
15	53	6.9	1.1
16	22	6.7	2.0

[a]Adapted from Dimock (1935).

Self-concept and self-esteem

Self-concept refers to perception of self, while self-esteem refers to the value placed on one's self-concept. In many studies, however, the terms are used interchangeably. Many early studies of the influence of physical activity focussed on self-concept as a global entity. Current emphases indicate different domains of the self-concept and changes in the domains from childhood through adolescence. Shavelson, Huber and Stanton (1976), for example, distinguish between academic and non-academic self-concept, the latter including social, emotional and physical domains. Each domain is further sub-divided into sub-areas which eventually relate to behaviour in specific situations. Of relevance to the influence of physical activity are the two sub-areas of the physical domain: physical ability and physical appearance.

Harter (1982, 1983, 1985, 1989; Harter & Pike, 1984) emphasize a developmental framework, focussing on changes in the 'self-system'. Based on a series of factor analyses, results suggest that children 4–7 years of age identify four domains: cognitive competence, physical competence, peer acceptance and maternal acceptance. Physical competence is described exclusively in the context of motor skills, e.g. 'good at climbing', 'good at running', 'good at hopping', etc. As noted earlier, these fundamental movement patterns are developing at these ages. Young children do not distinguish between cognitive and physical competence and between peer and maternal acceptance, resulting in two factors: general competence and social acceptance. Further, children do not apparently make judgements of their self-worth ('. . . conscious, verbalizable concept of one's worth as a person . . .' Harter, 1989, p. 70). Relevant to the discussion at hand, 4–7 year-old children do not distinguish between cognitive and physical skills: 'One is either competent or incompetent across these activities' (Harter, 1983, p. 331). Children do recognize, however, the importance of practice and learning for the development of competence.

The structure of self-concept changes at about 8 years of age, and has common domains through adolescence. Domains are, however, more clearly differentiated and children make judgements about self-worth. Among children 8–12 years of age, five domains emerge: scholastic competence, athletic competence, peer social acceptance, behavioural conduct, and physical appearance, while during adolescence three additional domains emerge: close friendship, romantic appeal, and job competence (Harter, 1989).

Studies of self-concept do not include actual measures of physical or athletic competence (i.e. skill) nor do they include measures of size, physique, maturity status, and so on. The relationship between perceived

Table 8.6. *Mean performances of children in grades K through 4 grouped on the basis of perceived competence*[a]

Test item	Perceived competence: Tertiles		
	Upper	Middle	Lower
MOTOR ABILITY ITEMS			
Standing long jump (cm)	120.9	117.6	108.4
Shuttle run – 60 yards (s)	18.1	18.5	18.6
Situps (number/30 s)	14.6	13.7	12.8
Sidestep (number of lines/30 s)	15.1	14.6	14.5
Flexed arm hang (s)	9.5	8.6	8.0
SPORT SKILLS			
Soccer dribble (number of cones/30 s)	7.2	6.5	5.9
Playground ball dribble (number of cones/30 s)	16.0	14.4	12.7
Softball repeated throws (number of throws/30 s)	10.3	9.7	9.0
Soccer ball throw (m)	5.0	4.4	4.0

[a]Adapted from Ulrich (1987).

and actual competence thus merits examination. In one of the few studies addressing this issue, Ulrich (1987) considered the relationship between perceived competence (assessed by Harter's scales) and demonstrated competence in basic motor abilities and sports skills in 250 children, kindergarten through grade 4 (mean ages by grade 5.8, 6.9, 7.9, 9.0, and 9.9 years). Results are summarized in Table 8.6. Children in the lowest tertile of perceived competence, on average, do not perform as well as children in the other tertiles in basic motor ability tasks and more specialized sports skills. The trends in the mean values suggest a dose–response effect, i.e. children with the highest perceived competence tend to show better levels of motor competence, more so in specialized sport skills than in basic motor ability items. The results suggest that 5–10 year-old children have a reasonably accurate perception of their motor competence and emphasize the need for developmentally appropriate programmes of physical education at these ages when basic motor skills are developing. Lack of success in motor activities at young ages may, in turn, have a significant impact on the motivation to participate in physical activities at later ages. This is relevant for individuals are often labelled early in life, and '. . . the poorly skilled are made aware quickly of their lack of skill' (Ulrich, 1987, p. 65).

Generally similar results are apparent at older ages. Youngsters involved in sport demonstrate greater levels of perceived physical competence compared to those who are not involved in youth sport (Roberts, Kleiber & Duda, 1981; Feltz & Petlichkoff, 1983). Further, currently active young athletes and those with more experience have higher levels of perceived physical competence than those who have dropped out and who have less experience in sport.

Results of these studies, though interesting, raise several questions. For example, what are the determinants of perceived competence? Do motor activities and sport attract children with high levels of perceived physical competence more so than other activities? Do children with lower levels of perceived competence shy away from motor activities? Are they the children who drop out of sport? Are higher levels of perceived competence a result of participation and/or success in motor activities and sport?

In contrast to the preceding observations, earlier data for children in grades 1–5 indicate no relationship between body control tasks (Body Coordination test of Schilling and Kiphard) and self-concept (Martinek-Zaichkowsky Self-Concept Scale), $r = 0.08$ and -0.1 before and after specific instructional programmes respectively (Martinek, Zaichkowsky & Cheffers, 1977). The Body Coordination test, however, is designed to diagnose children with motor problems so that it may not differentiate motor skill sufficiently in normal children. It includes backwards walking on a low balance beam, one foot hopping over blocks, side-to-side jumping and box placing, tasks which are passed by more than 75% of children 5–8 years of age (Kiphard & Schilling, 1970; Schilling & Kiphard, 1976).

The contributions of specific domains or competencies to self-worth also have implications for physical activity. Among elementary and middle school children (8–15 years), first physical appearance and then social acceptance have the most impact on children's self-worth, while more specific competencies, i.e. scholastic and athletic, contribute relatively little to self-worth at these ages (Harter, 1989). Thus, physical appearance and social acceptability have more impact on a child's sense of self-worth than specific competencies during middle childhood and early adolescence. This is a period of dramatic change in size, physique and body composition and in sexual maturity status for most children, so that it would seem logical to include more direct measures of these biological parameters in assessments of self-concept. Among children in grades 4–6, for example, body build indirectly assessed by the ponderal index is related to self-concept (Piers-Harris Children's Self-Concept Scale), and the relationship may change with age and may be different between boys and girls (Guyot, Fairchild & Hill, 1981). Among boys in grade 6 (about 12 years), self-concept (Staffieri's scale) is related to the ponderal index, those in the mid-range of the scores having a higher self-concept than those at either extreme. The same results, however, are not apparent in 9th grade boys (about 15 years). Self-concept scores do not differ among the three build groups; interestingly, the most linear boys had the highest self-concept scores (Felker, 1968). The results thus suggest that the influence of body build on self-concept declines with age during male adolescence (see also Felker & Kay, 1971). Note, however, that the ponderal index is only a proxy for physique based on the relationship of weight and stature.

Given concern for physical appearance among children and youth, regular physical activity has the potential to contribute to self-concept or self-worth. Although regular physical activity has a negligible influence on stature and physique, it can influence body composition, specifically fatness (Malina, 1990). It would be of interest to examine the print and electronic media, in addition to role models in the sport and entertainment industries, as determinants of this concern for physical appearance. Images of extreme slenderness and leanness, among others related to health habits, dress, hair style, etc, are aimed regularly at children and youth who are among the largest consumers of these media messages.

The literature dealing with the influence of physical activity programmes on self-concept/self-esteem is reasonably extensive. Results of a meta-analysis of 27 experimental studies generally indicate a favourable effect of physical activity on the self-concept of children and youth (Gruber, 1986; see Vogel, 1986, for a discussion of physical education programmes). Physical fitness and aerobic activity programmes have a greater influence (effect size = 0.89) than programmes of sport skills (0.40), creative dance (0.32) and perceptual-motor skills (0.29). There is no clear pattern of sex differences, but some evidence indicates greater gains in younger children. The programmes for young children generally focus on perceptual–motor and motor skill development, which are important developmental tasks at these ages. Physical fitness programmes have a similar positive influence on the self-esteem of adults (Sonstroem, 1984).

A shortcoming of experimental studies of physical activity and self-concept is that the persistence of activity-related effects is not ordinarily considered. Are the changes permanent? And, in the case of youth, do the changes associated with activity programmes interact with normal processes of growth and maturation? Can activity-related changes in self-concept be partitioned from those which accompany normal growth and maturation during the adolescent spurt, given the changes in physical appearance that occur?

Social competence

Physical activities ordinarily occur within a social context so that they have potentially significant implications for the development of social competence. Physical activities, however, do not socialize *per se*. They provide the medium in which the socialization process may occur. This process has three primary components: socializing agents or significant others, social environments or situations, and role learners (McPherson & Brown, 1988).

Socialization occurs *into* or *via* specific social roles. In the former, the individual is socialized into a specific role, i.e. a physically active or inactive lifestyle, or an athlete or a spectator. In the latter, the individual is socialized through specific roles into the learning of more general attitudes, values, and so on.

It is in the context of socialization *through* physical activity that the presumed social benefits of participation are often discussed. A variety of approaches have been used to assess the potential socialization benefits of physical activity, most often in the context of sport. Given the site of this symposium, it is noteworthy that games and team sports in the school setting emerged in mid-19th century England. They were justified in a largely social context, i.e. as a primary avenue for character training. According to the Royal Commission on Public Schools in 1864,

> The bodily training which gives health and activity to the frame is imparted at English schools, not by gymnastic exercises which are employed for that end on the continent – exercises which are undoubtedly very valuable and which we would be glad to see introduced more widely in England – but by athletic games which, whilst they serve this purpose well, serve other purposes besides . . . the cricket and football fields . . . are not merely places of exercise or amusement; they help to form some of the most valuable social qualities and manly virtues, and they hold, like the classroom and the boarding house, a distinct and important place in Public School Education (quoted in McIntosh, 1981, pp. 185–6).

Historically school games and sports were organized and administered by students. Today, however, sports for children and youth are adult-governed enterprises, and character development, i.e. 'learning socially acceptable values and behaviours', is an expressed objective, among others, of youth sports programmes (Seefeldt, 1987).

Most physical activities, including sport, provide a variety of social experiences or interactions. The question of interest, of course, is whether these experiences contribute to the development of socially competent behaviour. Definitions of social competence vary (see McFall, 1982; Trower, 1982; Dodge *et al.*, 1986).

Systematic study of the effects of participation in physical activities, including sport, on social development or the development of social competence is rather limited. The literature tends to focus on specific social behaviours, values and characteristics of those who are proficient in motor skills or those who are involved in sport. It is often argued, for example, that the child who is proficient in motor skills usually possesses social status among his/her peers. In the world of children and youth, the peer group is central, and physical activities occur within the framework of the group and/or through peer sanction and support. Children generally have multiple peer groups (e.g. school, church, team, sport club, neighbourhood, etc), each with its own social demands and experiences.

In one of the earlier studies, third grade children (5 boys, 5 girls, about 8 years) proficient in motor skills, i.e. physically competent, '. . . tended to be active, popular, calm, resourceful, attentive, and cooperative . . .' compared to a corresponding sample of chidren who were not proficient (Rarick & McKee, 1949, p. 150). Other evidence suggests that elementary school children competent in motor skills attain greater social success and status than the less competent, and that leadership and peer acceptance is related to proficiency in motor skills (Evans & Roberts, 1987). In the context of youth sports, the internal structure of the team sport and particularly skill may be important mediating factors in peer relations (Bigelow, Lewko & Salhani, 1989), while fourth and fifth grade boys and girls (about 9–10 years) involved in sport are rated higher in social competence than those not involved in sport (Roberts *et al*., 1981).

During the transition into adolescence, variation in biological maturity status enters the matrix of factors which influences social status and the socialization process. Individual differences in the timing of biological maturation and associated changes in size and body composition are a major component of the backdrop against which children evaluate and interpret their social status among peers. Physical performance and success in sports is an important aspect of the evaluative process, particularly among males, and biological maturation has significant correlates in physical skills and the value attached to skills by the peer group. Advanced (early) maturation among boys is associated with proficiency in strength and motor performance, as well as with popularity and social status (Jones, 1949). Early maturation in boys is also associated with success in many youth sports (Malina, 1988), with resulting advantages in social status.

The situation is different for girls. Differences in strength and performance among girls of contrasting maturity status are not as apparent as among boys during the transition into adolescence. Further, it is often the late maturing girl who performs better and who experiences success in sports and persists in sports through the transition from childhood into adolescence (Malina, 1988). This raises two questions (and undoubtedly many others) which merit systematic study. Are late maturing girls socialized into sport, or are early maturing girls socialized away from sport?

Data on the socialization of young girls into sport and on the social status of young female athletes are lacking (see Lewko & Greendorfer, 1988). The status of the young teenage girl in her social group is often linked to her femininity, and until recently, sport was not generally considered feminine. With the increased acceptance of women as athletes and with more opportunities for young girls to participate in sport, this perception is changing. Many pre-adolescent and adolescent boys accept and value female peers who are skilled in sport, given the opportunity to do so. This is one of the major benefits of co-educational youth sports programmes.

Cognitive competence

The relationship of physical activity and cognitive competence has been approached primarily in the context of intellectual development and academic achievement. A good deal of the academic achievement literature deals with comparisons for athletes and non-athletes, comparisons which involve too many uncontrolled factors (e.g. athletes must maintain a minimum grade to participate). Other literature takes two directions, the first correlating motor skill or physical activity with intellectual performance, and the second being experimental.

The association between motor performance and intellectual development is more apparent at the extremes of skill and intelligence. Several examples serve to illustrate the kinds of studies and trends in results. Third grade children with high motor proficiency in tests of running, jumping, throwing, striking, catching, agility and balance ($n = 10$ per group, 5 boys, 5 girls) have a higher IQ and excellent/good ratings in reading, writing and comprehension compared to those with low motor proficiency (Rarick & McKee, 1949). In a similar study of 9, 12 and 15 year-old boys equated for IQ ($n = 20$ per age and strength group), high in a strength index and strength adjusted for age and weight tend to score higher on tests of academic achievement and school grades than boys low in strength (Clarke & Jarman, 1961). Finally, comparison of highly active ('great amount of participation in sports, games, and other vigorous physical recreational activities in or out of school in all seasons of the year') and inactive ('little or no such participation') boys equally distributed between grades 7 and 12 ($n = 50$ per group, mean age 15.7 years) indicates significantly higher cumulative and current grades in the active group (Ruffer, 1965).

In an interesting series of studies, Ismail and colleagues (Ismail, 1964; Ismail & Gruber, 1967; Ismail, Kane & Kirkendall, 1969; see also Kirkendall, 1986) attempted to predict IQ and academic achievement of fifth and sixth grade children (10–12 years) from performance on a large series of motor tasks. Coordination tasks followed by balance tasks tend to be the best predictors of IQ and academic achievement, while items of strength, speed and power contribute little. Coordination and balance items also differentiate high and low achieving children, the former performing significantly better in the majority of tasks. In contrast, the two groups do not differ in tasks requiring strength, power and speed.

The literature thus suggests a positive relationship between proficiency in motor skills and cognitive competence as measured by IQ and academic achievement. Few studies address the issue of physical activity *per se*, but active children tend to be more proficient in motor skills. The magnitude of the relationship varies from low to moderate, and in studies utilizing more extensive motor test batteries, coordination and balance items are more related to intellectual performance. Correlation, of course, does not imply a cause–effect sequence of events.

Experimental studies are fewer in number and ordinarily focus on the effects of specific physical education programmes, or additional instruction in physical education (see Vogel, 1986). Ismail and Gruber (1967), for example, considered the influence of an organized physical education programme on the IQ and academic achievement of fifth and sixth graders. The design controlled for age, sex, level of achievement, IQ, and pairing of subjects, who were divided into experimental and control groups. The experimental physical education programme had no influence on IQ, but influenced academic achievement scores favourably.

Perceptual motor training programmes are related in part to the preceding discussion. Such programmes ordinarily focus on training in gross and fine motor skills identified as relevant to academic readiness or performance, especially of children with learning disabilities, e.g. eye–hand coordination, dynamic balance, directionality of movement, and so on. Results of numerous studies, however, are equivocal. The learning of motor skills is rather task specific and does not generalize to cognitive performance (for reviews, see Seefeldt, 1974; Rarick, 1980; Thomas & Thomas, 1986).

It should not be overlooked that physical activities are routinely used in the teaching of cognitive skills. Manipulative activities, both fine and gross, are commonly used in teaching mathematics (Burton, 1984; del Grande, 1990). Spatial visualization and estimation, seriation (pattern recognition and development), and measurement, for example, are developed with physical manipulation, especially in primary grade children. Sample activities might require a child to estimate how far he/she can jump in three jumps, or how many boxes of a given size will fill a room.

Motor activities are also used in the development of memory for spatial location, and they are more effective with younger children. In an unfamiliar environment, for example, 5 year-old children depend more on motor activity for the development of location accuracy than do 8 year-old children (Figure 8.1). The placement accuracy of younger children

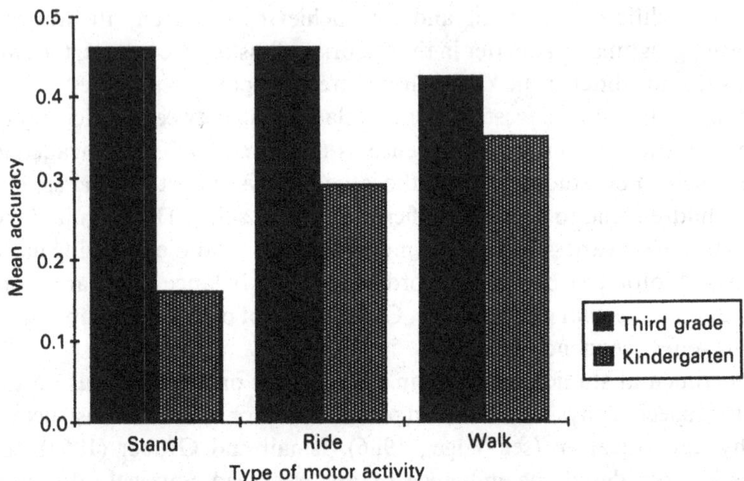

Figure 8.1. Mean placement accuracy (memory for spatial location) in kindergarten and third grade children as a function of the amount of motor involvement in the environment. (Redrawn after Herman, Kolker & Shaw, 1982.)

improves linearly from standing (stand and survey a town model about $5 \times 6\,m^2$), to riding (sit in a wagon and be pulled through the model), to walking (walk through the model). In contrast, the placement accuracy of 8 year-old children does not differ among the three types of motor activity (Herman, Kolker & Shaw, 1982). The preceding examples thus suggest that motor activities and specific skills may become a pathway to learning cognitive concepts.

Summary

Physical activities are an important component of behavioural development. Influences undoubtedly vary with age and perhaps sex, and of course, with the context of the activities. The systematic study of physical activity has not been a major objective in human biology research. The time is ripe, nevertheless, for physical activity to assume a central position in the biocultural framework of human biology.

References

American Alliance for Health, Physical Education, and Recreation (1976). *AAHPER Youth Fitness Test Manual*. Reston, Va.: American Association for Health, Physical Education and Recreation.

American Alliance for Health, Physical Education, Recreation, and Dance (1980).

Health Related Physical Fitness Test Manual. Reston, Va.: American Alliance for Health, Physical Education, Recreation, and Dance.

American Alliance for Health, Physical Education, Recreation, and Dance (1984). *Technical Manual: Health Related Physical Fitness.* Reston, Va.: American Alliance for Health, Physical Education, Recreation, and Dance.

Beunen, G. & Malina, R.M. (1988). Growth and physical performance relative to the timing of the adolescent spurt. *Exercise and Sport Sciences Reviews*, **16**, 503–40.

Beunen, G.P., Malina, R.M., Van't Hof, M.A., Simons, J., Ostyn, M., Renson, R. & Van Gerven, D. (1988). *Adolescent Growth and Motor Performance: A Longitudinal Study of Belgian Boys.* Champaign, Il.: Human Kinetics Books.

Bigelow, B.J., Lewko, J.H. & Salhani, L. (1989). Sport involved children's friendship expectations. *Sport and Exercise Psychology*, **11**, 152–60.

Bouchard, C., Shephard, R.J., Stephens, T., Sutton, J.R. & McPherson, B.D. (1990). Exercise, fitness, and health: the consensus statement. In *Exercise, Fitness, and Health.* Bouchard, C., Shephard, R.J., Stephens, T., Sutton, J.R. and McPherson, B.D., eds, pp. 3–31. Champaign, Il.: Human Kinetics Books.

Burton, L. (1984). Mathematical thinking: the struggle for meaning. *Journal for Research in Mathematics Education*, **15**, 35–49.

Canadian Association for Health, Physical Education, and Recreation (1980). *The CAHPER Fitness-Performance Test Manual.* Ottawa: Canadian Association for Health, Physical Education, and Recreation.

Clarke, H.H. & Jarman, B.O. (1961). Scholastic achievement of boys 9, 12, and 15 years of age as related to various strength and growth measures. *Research Quarterly*, **32**, 155–62.

Dimock, H.S. (1935). A research in adolescence: the social world of the adolescent. *Child Development*, **6**, 285–302.

Dodge, K.A., Pettit, G.S., McClaskey, C.L. & Brown, M.M. (1986). Social competence in children. *Monographs of the Society for Research in Child Development*, **51**, No. 213.

Evans, J. & Roberts, G.C. (1987). Physical competence and the development of children's peer relations. *Quest*, **39**, 23–35.

Felker, D.W. (1968). Relationship between self-concept, body build, and perception of father's interest in sport in boys. *Research Quarterly*, **39**, 513–23.

Felker, D.W. & Kay, R.S. (1971). Self-concept, sports interests, sports participation, and body type of seventh- and eighth-grade boys. *Journal of Psychology*, **78**, 223–8.

Feltz, D. & Petlichkoff, L. (1983). Perceived competence among interscholastic sport participants and dropouts. *Canadian Journal of Applied Sport Sciences*, **8**, 231–5.

Gould, D. & Horn, T. (1984). Participation motivation in young athletes. In *Psychological Foundations of Sport*, Silva, J.M. and Weinberg, R.S., eds, pp. 359–70. Champaign, Il.: Human Kinetics Publishers.

del Grande, J. (1990). Spatial sense. *Arithmetic Teacher*, **37**, 14–20. (February).

Gruber, J.J. (1986). Physical activity and self-esteem development in children: a meta-analysis. In *Effects of Physical Activity on Children* (American Academy of Physical Education Papers No. 19), Stull, G.A. and Eckert, H.M., eds, pp. 40–8. Champaign, Il.: Human Kinetics Publishers.

Guyot, G.W., Fairchild, L. & Hill, M. (1981). Physical fitness, sport participation, body build, and self-concept of elementary school children. *International*

Journal of Sport Psychology, **12**, 105–16.

Harter, S. (1982). The perceived competence scale for children. *Child Development*, **53**, 87–97.

Harter, S. (1983). Developmental perspectives on the self-esteem. In *Handbook of Child Psychology, vol. IV. Socialization, Personality, and Social Development*. Hetherington, E.M., ed., pp. 275–385. New York: Wiley.

Harter, S. (1985). Competence as a dimension of self-evaluation: Toward a comprehensive model of self-worth. In *The Development of the Self*. Leahy, R.L., ed., pp. 55–121. New York: Academic Press.

Harter, S. (1989). Causes, correlates, and the functional role of global self-worth: a life-span perspective. In *Perceptions of Competence and Incompetence across the Life-Span*. Kolligian, J. and Sternberg, R., eds, pp. 67–97. New Haven, Ct.: Yale University Press.

Harter, S. & Pike, R. (1984). The pictorial scale of perceived competence and social acceptance for young children. *Child Development*, **55**, 1969–82.

Haubenstricker, J.L. & Seefeldt, V.D. (1986). Acquisition of motor skills during childhood. In *Physical Activity and Well-Being*. Seefeldt, V., ed., pp. 41–102. Reston, Va.: American Alliance for Health, Physical Education, Recreation, and Dance.

Herman, J.F., Kolker, R.G. & Shaw, M.L. (1982). Effects of motor activity on children's intentional and incidental memory for spatial locations. *Child Development*, **53**, 239–44.

Ismail, A.H. (1964). Motor aptitude items and academic achievement. In *Report of a Symposium on Integrated Development*, pp. 19–31. Lafayette, In.: Purdue University.

Ismail, A.H. & Gruber, J.J. (1967). *Integrated Development: Motor Aptitude and Intellectual Performance*. Columbus, Oh.: Charles E. Merrill Books.

Ismail, A.H., Kane, J. & Kirkendall, D.R. (1969). Relationships among intellectual and nonintellectual variables. *Research Quarterly*, **40**, 83–92.

Jones, H.E. (1949). *Motor Performance and Growth*. Berkeley: University of California Press.

Kemper, H.C.G., Dekker, H.J.P., Ootjers, M.G., Post, B., Snel, J., Splinter, P.G., Storm-van Essen, L. & Verschuur, R. (1983). Growth and health of teenagers in the Netherlands. *International Journal of Sports Medicine*, **4**, 202–14.

Kiphard, E.J. & Schilling, F. (1970). Der Hamm-Marburger Korperkoordinationstest fur Kinder (HMKTK). *Monatsschrift fur Kinderheilkunde*, **118**, 473–9.

Kirkendall, D.R. (1986). Effects of physical activity on intellectual development and academic performance. In *Effects of Physical Activity on Children* (American Academy of Physical Education Papers No. 19). Stull, G.A. and Eckert, H.M., eds, pp. 49–63. Champaign, Il.: Human Kinetics Publishers.

Lewko, J.H. & Greendorfer, S.L. (1988). Family influences in sport socialization of children and adolescents. In *Children in Sport*, (3rd edn). Smoll, F.L., Magill, R.A. and Ash, M.J., eds, pp. 287–300. Champaign, Il.: Human Kinetics Books.

McFall, R.M. (1982). A review and reformulation of the concept of social skills. *Behavioral Assessment*, **4**, 1–35.

McIntosh, P.C. (1981). Games and gymnastics for two nations in one. In *Landmarks in the History of Physical Education*. McIntosh, P.C., Dixon, J.G., Munrow, A.D. and Willetts, R.F., eds, pp. 185–217. London: Routledge, Kegan & Paul.

McPherson, B.D. & Brown, B.A. (1988). The structure, processes, and consequences of sport for children. In *Children in Sport*, (3rd edn). Smoll, F.L., Magill, R.A. and Ash, M.J., eds, pp. 265–86. Champaign, Il.: Human Kinetics Books.

Malina, R.M. (1986). Energy expenditure and physical activity during childhood and youth. In *Human Growth: A Multidisciplinary Review*, Demirjian, A., ed., pp. 215–25. London: Taylor & Francis.

Malina, R.M. (1988). Growth and maturation of young athletes: Biological and social considerations. In *Children in Sport*, (3rd edn). Smoll, F.L., Magill, R.A. and Ash, M.J., eds, pp. 83–101. Champaign, Ill.: Human Kinetics Books.

Malina, R.M. (1990). Growth, exercise, fitness, and later health outcomes. In *Exercise, Fitness, and Health.* Bouchard, C., Shephard, R.J., Stephens, T., Sutton, J.R. and McPherson, B.D., eds, pp. 637–65. Champaign, Il.: Human Kinetics Books.

Malina, R.M. (1991). Fitness and performance: Adult health and the culture of youth. In *New Possibilities, New Paradigms?* (American Academy of Physical Education Papers No. 24). Park, R.J. and Eckert, H.M., eds, pp. 30–8. Champaign, Il.: Human Kinetics Publishers.

Malina, R.M. & Bouchard, C. (1991). *Growth, Maturation, and Physical Activity.* Champaign, Il.: Human Kinetics Books.

Martinek, T.J, Zaichkowsky, L.D. & Cheffers, J.T.F. (1977). Decision-making in elementary age children: effects on motor skills and self-concept. *Research Quarterly*, **48**, 349–57.

Mirwald, R.L. & Bailey, D.A. (1986). *Maximal Aerobic Power.* London, Ontario: Sports Dynamics.

Moravec, R. (1989). Vplyv vykonavanych pohybovych aktivit na telesny rozvoj a pohybovu vykonnost 7–18-rocnej mladeze. *Teorie a Praxe Telesny Vychovy*, **37**, 596–606.

Parizkova, J. (1977). *Body Fat and Physical Fitness.* The Hague: Martinus Nijhoff.

Pate, R.P. & Ross, J.G. (1987). Factors associated with health-related fitness. *Journal of Physical Education and Recreation*, **58**, 93–5 (November–December).

Pate, R.R., Dowda, M. & Ross, J.G. (1990). Associations between physical activity and physical fitness in American children. *American Journal of Diseases of Children*, **144**, 1123–9.

Rarick, G.L. (1980). Cognitive–motor relationships in the growing years. *Research Quarterly*, **51**, 174–92.

Rarick, G.L. & McKee, R. (1949). A study of twenty third-grade children exhibiting extreme levels of achievement on tests of motor proficiency. *Research Quarterly*, **20**, 142–52.

Roberts, G.C., Kleiber, D.A. & Duda, J.L. (1981). An analysis of motivation in children's sport: the role of perceived competence in participation. *Journal of Sport Psychology*, **3**, 206–16.

Ross, J.G., Dotson, C.O., Gilbert, G.G. & Katz, S.J. (1985). After physical education . . . physical activity outside of school physical education programs. *Journal of Physical Education and Recreation and Dance*, **56**, 77–81 (January).

Ruffer, W.A. (1965). A study of extreme physical activity groups of young men. *Research Quarterly*, **36**, 183–96.

Sapp, M. & Haubenstricker, J. (1978). Motivation for joining and reasons for not continuing in youth sports programs in Michigan. Paper presented at the

Annual Convention of the American Alliance for Health, Physical Education and Recreation, Detroit, Michigan (cited by Seefeldt & Haubenstricker, 1988).

Schilling, F. & Kiphard, E.J. (1976). The body coordination test. *Journal of Physical Education and Recreation*, **47**, 37 (April).

Schmucker, B., Rigauer, B., Hinrichs, W. & Trawinski, J. (1984). Motor abilities and habitual physical activity in children. In *Children and Sport*. Ilmarinen, J. and Valimaki, I., eds, pp. 46–52. Berlin: Springer-Verlag.

Seefeldt, V. (1974). Perceptual–motor programs. *Exercise and Sports Sciences Reviews*, **2**, 265–88.

Seefeldt, V. (1987). Benefits of competitive sports for children and youth. In *Handbook for Youth Sports Coaches*. Seefeldt, V., ed., pp. 3–15. Reston, Va.: American Alliance for Health, Physical Education, Recreation, and Dance.

Seefeldt, V. & Haubenstricker, J. (1982). Patterns, phases, or stages: An analytical model for the study of developmental movement. In *The Development of Movement Control and Co-ordination*. Kelson, J.A.S. and Clark, J.E., eds, pp. 309–19. New York: Wiley.

Seefeldt, V. & Haubenstricker, J. (1988). Children and youth in physical activity: A review of accomplishments. In *Physical Activity in Early and Modern Populations* (American Academy of Physical Education Papers No. 21). Malina, R.M., ed., pp. 88–103. Champaign, Il.: Human Kinetics Books.

Shavelson, R.J., Huber, J.J. & Stanton, G.C. (1976). Self-concept: validation of construct reinterpretation. *Review of Educational Research*, **46**, 407–41.

Simons, J. & Renson, R. (eds) (1982). *Evaluation of Motor Fitness*. Leuven, Belgium: Katholieke Universiteit Leuven, Institute of Physical Education.

Sonstroem, R.J. (1984). Exercise and self-esteem. *Exercise and Sport Sciences Reviews*, **12**, 123–55.

Telama, R. & Silvennoinen, M. (1979). Structure and development of 11 to 19 year olds' motivation for physical activity. *Scandinavian Journal of Sport Sciences*, **1**, 23–31.

Thomas, J.R. & Thomas, K.T. (1986). The relation of movement and cognitive function. In *Physical Activity and Well-Being*. Seefeldt, V., ed., pp. 443–52. Reston, Va.: American Alliance for Health, Physical Education, Recreation, and Dance.

Trower, P. (1982). Toward a generative model of social skills: a critique and synthesis. In *Social Skills Training*. Curran, J.P. and Monti, P.M., eds, pp. 399–428. New York: Guilford.

Ulrich, B.D. (1987). Perceptions of physical competence, motor competence, and participation in organized sport: their interrelationships in young children. *Research Quarterly for Exercise and Sport*, **58**, 57–67.

Vogel, P.G. (1986). Effects of physical education programs on children. In *Physical Activity and Well-Being*, Seefeldt, V., ed., pp. 455–509. Reston, Va.: American Alliance for Health, Physical Education, Recreation, and Dance.

Watson, A.W.S. & O'Donovan, D.J. (1977). The relationship of level of habitual activity to measures of leanness–fatness, physical working capacity, strength and motor ability in 17 and 18 year-old males. *European Journal of Applied Physiology*, **37**, 93–100.

Wellens, R.E. (1989). Activity as a temperamental trait: relationship to physique, energy expenditure and physical activity habits in young adults. PhD Thesis. University of Texas at Austin.

9 *Physiological aspects of activity and ageing*

E.J. BASSEY

Old age is associated with increased morbidity which is well documented; it is also associated with a decrease in customary physical activity which is less well described. The assessment of customary activity levels is not easy even in young individuals many of whom participate regularly in sport; it is much more difficult in old individuals for whom walking is usually their most active pursuit.

There are a number of approaches which have been reviewed (Patrick, 1985). They range from direct methods using simple postal questionnaires to indirect methods which depend on logging bio-signals which are related to activity level such as heart rate or foot-fall (Patrick *et al.*, 1986). Questionnaires may be of dubious validity in the elderly since they often have poor short-term recall, and there is no single bio-signal which has good predictive power for a variety of different activities. There is no ideal approach despite some sophisticated techniques. Assessment of metabolic rate using labelled water promises to provide the best method but awaits validation for old people. Customary activity therefore has to be described within the limitations of the measurement technique.

Retirement from full-time employment marks a change in lifestyle which may be associated with changes in activity level. It is a potential watershed for a reduction in activity which could lead to declining function and so to further reductions in customary activity, forming a vicious circle of deterioration. Retirement studies were therefore conducted on small numbers of male and female factory workers using both structured questionnaires and objective measurements.

Men retiring from a steelworks at 65 years of age were assessed in the 3 months before retirement and then annually for several years (Patrick *et al.*, 1986). Body borne tape recorders were used to record heart rates and footfall signals during the whole of the waking day. This data was used to construct activity indices based on both duration and intensity of activity. Arbitrary, but rational, thresholds were set, for instance, for time spent walking from the footfall signal and for exercise intensity from heart rate level. The heart rates were calibrated individually for each man and the

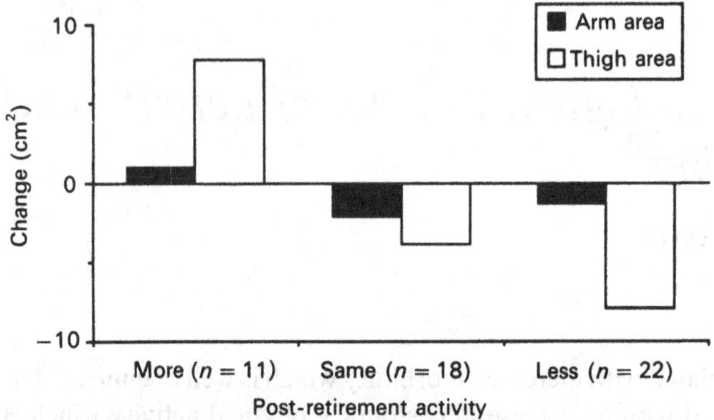

Figure 9.1. Changes in muscle cross-sectional area according to reported change in physical activity level after 1 year of retirement in 51 male steelworkers aged 65 years.

thresholds were set at the heart rate required to walk at 3 mph (4 km/h). This is a normal walking speed for a healthy adult. Measurements of body composition were also made, derived from four skinfolds and also a soft tissue radiograph of the thigh (Patrick, Bassey & Fentem, 1982).

The outcome was that very little activity was recorded even before retirement. Walking was rarely continued for more than 2 minutes and heart rates seldom rose above the threshold level for long. However, during the first year of retirement a substantial number of individuals reported increased activity levels and those who reported increased activity were found to have an increased amount of muscle (see Figure 9.1). Few of these men had cars, many rode bicycles and some cultivated large allotment gardens. For some, retirement evidently brought time and opportunity for more physical activity rather than a decline.

Three years after retirement a significant decrease in one of four activity indices was observed in a small group of these men (Figure 9.2). The index to show a decrease was influenced by both intensity and duration of activity. Since the indices based only on duration of activity showed no change, and since the intensity threshold (heart rate required to walk at 3 mph) was always calibrated for each individual at each measurement, then these men must have been walking more slowly after retirement. Other factory workers were studied over a year pre- and post-retirement and women but not men showed a significant decrease in activity levels (see Figure 9.2).

Body composition measurements in rather larger numbers showed, on average, a significant increase in % fat during the first year of retirement in male steel workers and also in female factory workers. In the steel workers

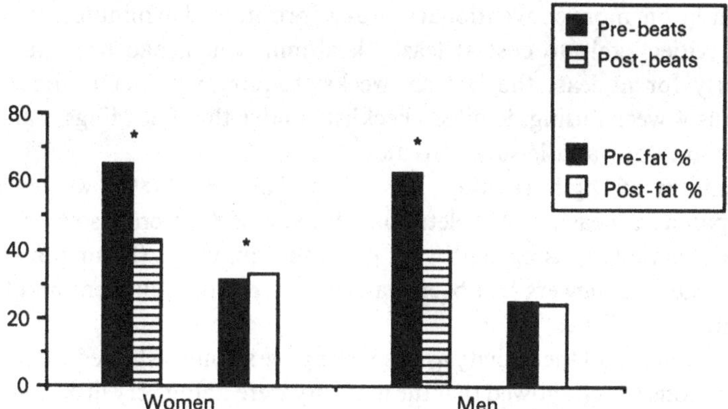

Figure 9.2. Changes after retirement in % body fat and physical activity level, the latter assessed over a whole day for cumulated continuous walking periods of at least 7.5 min. The activity index is cumulated time multiplied by the heart rates during the activity periods to give an intensity dimension. The units are therefore 'beats'. The filled columns represent data for 16 male steel workers (3 years after retirement) and the open columns 16 female workers from a light industry (1 year after retirement). Significant differences are identified (*$P<0.05$).

over the next 6 years of retirement, a slow fall in both fat and muscle occurred. These changes are consistent with the general view of reduced activity levels in retirement.

In 1985 a randomized survey of the physical and psychological well-being of 1042 men and women aged over 65 years and living independently in their own homes was conducted using a structured questionnaire (Dallosso *et al.*, 1988). The survey sample was drawn from an area which is demographically representative for the UK and stratified equally into two age-groups aged 65–74 years and over 75 years. The sample was balanced to match the sex distribution of the UK in these age groups and the response rate was 80%.

Physical activity was assessed in several ways. Since walking was likely to be the most frequent activity in this age-group, it was a major focus of the inquiry. However, because it is a commonplace and ubiquitous activity it is not memorable, moreover old people tend to have poor short-term recall. Therefore a detailed inquiry was made, with suitable prompts, about the walking events which had taken place on the previous day (provided that day had not been unusual). The total time spent walking outside the house and garden on that day was added up to provide a score. This probably gives the most accurate information for collating into group data; however it provides a poor sample for description of an individual, since walking activity will vary from day to day.

In addition, more conventional scores were obtained in minutes per week for activities likely to cost at least 2 kcal/min, which had been pursued regularly for at least the last six weeks. Inquiry was made about the previous 4 weeks using detailed checklists under three headings, namely, indoor, outdoor and leisure activities.

Handgrip strength (Bassey, Dudley & Harries, 1986), weight and demispan as a measure of skeletal size (Bassey, 1987) were also measured by the interviewers using calibrated portable equipment (Lehmann *et al.*, 1991). The interviewers had been trained to use this equipment and their reliability established.

The results of all the activity assessments gave strongly skewed frequency distributions which showed that the majority were doing very little and that a few were very active. Figure 9.3 shows the distribution of time spent

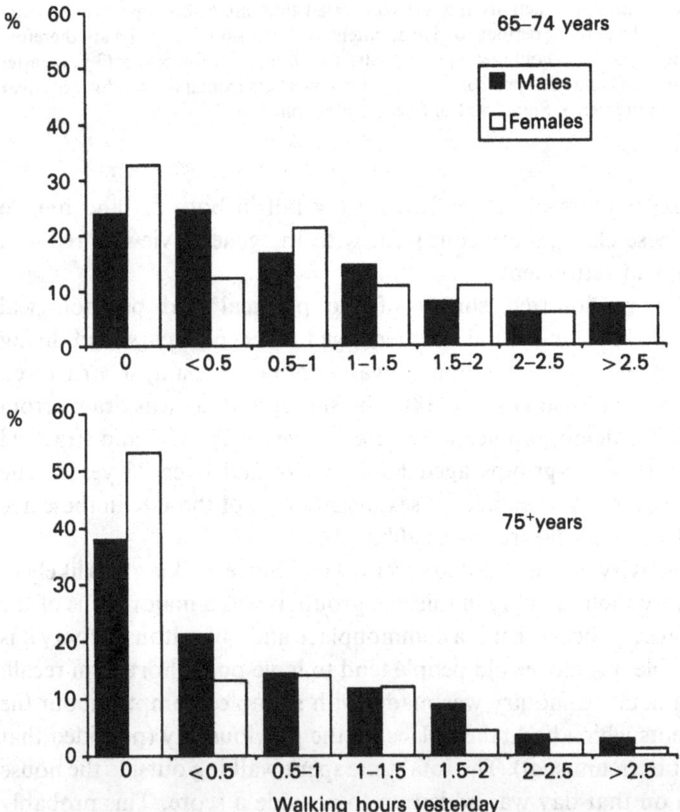

Figure 9.3. Frequency distribution (in percentages of each sub-group) of time spent walking yesterday (i.e. on the day previous to the interview, provided that day was not unusual). The younger sample is in the top half of the panel and the older in the lower half. The men are identified with filled columns and the women with open columns.

yesterday walking (excluding time spent shopping). This distribution changed little when those who were unable to walk outside the home and garden were omitted. The impression is that, on the whole, activity levels were low, although the survey was run in the summer months. Activity levels in the winter are likely to be even lower.

Leisure pursuits included social walking, keep-fit (at home), walking the dog, cycling, slow dancing and swimming. Social walking was the most prevalent activity with over 30% of the whole sample participating. Keep-fit was reported mainly by women. The other activities all had a prevalence well below 10%. Men spent more time than women in leisure pursuits just as in younger age groups (see Figure 9.4).

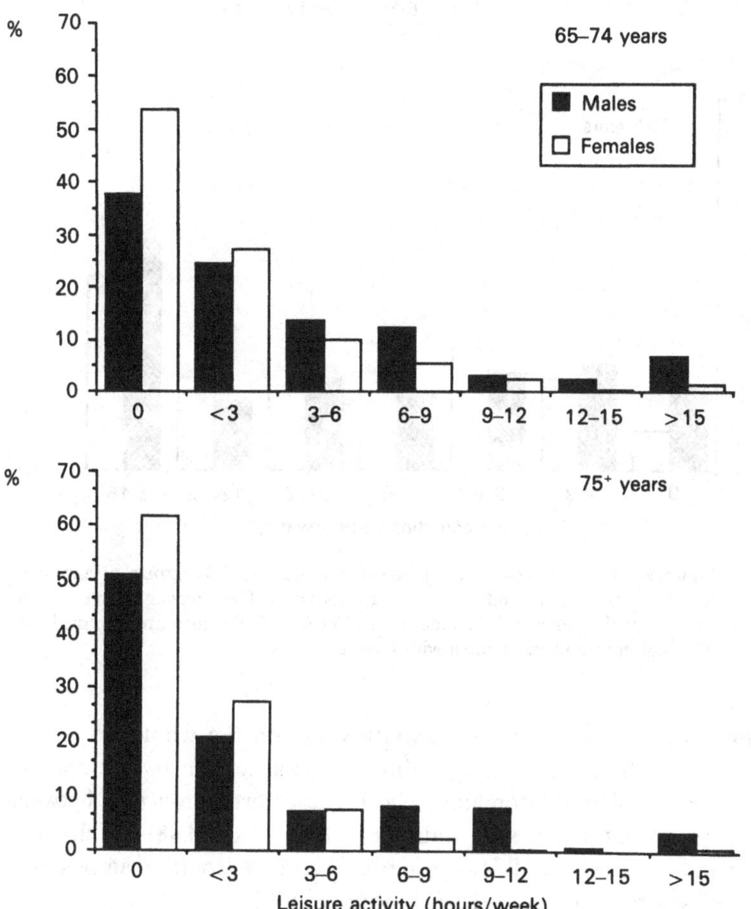

Figure 9.4. Frequency distribution (in percentages of each sub-group) of time spent in leisure activities. The younger sample is in the top half of the panel and the older in the lower half. The men are identified with filled columns and the women with open columns.

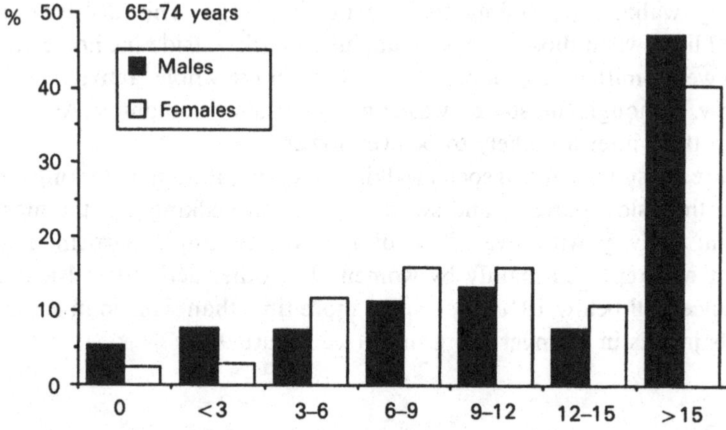

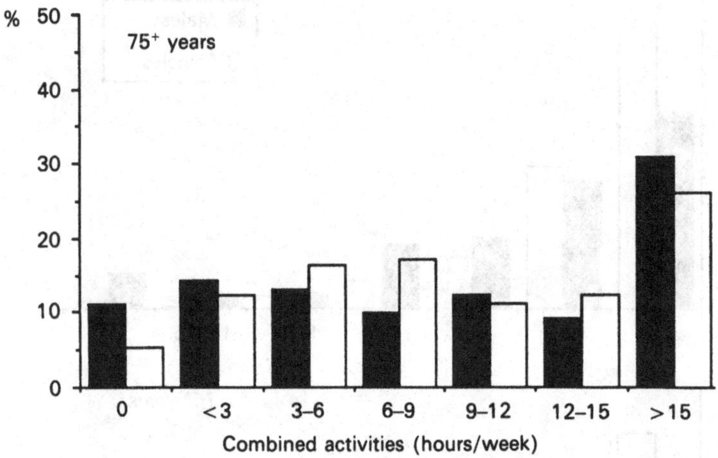

Figure 9.5. Frequency distribution (in percentages of each subgroup) of cumulated time spent in outdoor, indoor and leisure activities. The younger sample is in the top half of the panel and the older in the lower half. The men are identified with filled columns and the women with open columns.

These various assessments of activity were, on the whole, correlated significantly with each other, suggesting that, in general, active individuals were active in all these domains. The highest correlation was between walking yesterday and leisure walking in men (rho = 0.48) which lends some validity to the walking yesterday score. Also in men walking yesterday was not related to either indoor or outdoor activities or leisure activities as a whole. Other correlations were rather weak ranging from rho = 0.33 to 0.14. Factor analysis of all the variables which purport to measure customary activity has shown that, for women, there were two separable factors and for men there were three (Morgan *et al.*, 1991). These are dealt with fully in another chapter.

As expected, the men did more outdoor work than women and the women did more indoor work than men. When the three assessments (indoor, outdoor and leisure activities) were added together a more normal frequency distribution than for each activity separately was obtained and there was now little difference between the sexes (see Figure 9.5).

It is apparent from Figures 9.3–5, that activity declined with age in both men and women and this was significant except for indoor activity in men. This is presumably because the younger men were already doing very little of this.

Handgrip strength was significantly related to leisure walking in both men and women (rho = 0.15 and 0.14 respectively) and also to walking yesterday in men (rho = 0.16). Although this association is significant it is weak, explaining only 3% of the variance and it may appear because both variables are negatively related to age. Handgrip strength is significantly related to other muscle strengths and it may be a marker for health to which it is weakly related.

In 1989, four years after the original survey, a second survey was conducted and over 600 of the survivors were successfully re-interviewed. Longitudinal comparisons allowing realistic assessment of change with time in these age groups will be published in due course.

References

Bassey, E.J. (1987). Demi-span as a measure of skeletal size. *Annals of Human Biology,* **13**, 499–502.

Bassey, E.J., Dudley, B.R. & Harries, U.J. (1986). A new portable strain-gauge hand-grip dynamometer. *Journal of Physiology*, **373**: 6P.

Dallosso, H.M., Morgan, K., Bassey, E.J., Ebrahim, S., Fentem, P.H. & Arie, T.H.D. (1988). Levels of customary physical activity among the old and the very old living at home. *Journal of Epidemiology and Community Health*, **42**, 212–17.

Lehmann, A.B., Bassey, E.J., Morgan, K. & Dallosso, H.M. (1991). Normal values for weight, skeletal size and body mass indices in 890 men and women aged over 65 years. *Clinical Nutrition*, **10**, 18–22.

Morgan, K., Dallosso, H.M., Bassey, E.J. & Fentem, P.H. (1991). Customary physical activity and psychological well-being among elderly people living at home. In *Sport, Health, Psychology and Exercise*, Biddle, S., ed., pp. 135–46. London: Sports Council & Health Education Authority.

Patrick, J.M. (1985). Customary physical activity in the elderly. In *The Biology of Human Ageing*, Bittles, A.H. and Collins, K.J., eds, pp. 242–59. Cambridge: Cambridge University Press.

Patrick, J.M., Bassey, E.J. & Fentem, P.H. (1982). Changes in body fat and muscle in manual workers at and after retirement. *European Journal of Applied Physiology*, **49**, 187–96.
Patrick, J.M., Bassey, E.J., Irving, J.M., Blecher, A. & Fentem, P.H. (1986). Objective measurements of customary physical activity in elderly men and women before retirement. *Quarterly Journal of Experimental Physiology*, **71**, 47–58.

10 *Activity and morale in later life: preliminary analyses from the Nottingham Longitudinal Study of Activity and Ageing*

KATE BENNETT AND KEVIN MORGAN

Physical activity has been associated with a variety of cognitive and behavioural changes which, collectively, allow the conclusion that regular exercise contributes to psychological as well as physical health (Folkins & Sime, 1981; Veale, 1987). Thus, while concern has not moved away entirely from *whether* certain types of exercise have psychological benefits (e.g. Hughes, 1984) considerable attention is now being paid to how these benefits are mediated, and at whom they might best be targeted. Retired and elderly people in particular are being seen increasingly as appropriate candidates for health promotion initiatives which, by increasing levels of physical activity (and, by implication, levels of physical fitness), aim broadly to improve quality of life.

In this context, levels of customary or habitual activity (as distinct from levels of formal exercise participation) are presumed to play an important part, and are now beginning to receive research attention (Shephard & Montelpare, 1988). At present, however, the empirical basis for activity-based health promotion initiatives among elderly people owes much to information derived from younger age groups. Certainly, as regards mental health, relatively little research has directly addressed the assumption that customary physical activity reliably contributes to psychological well-being in later life.

The Nottingham Longitudinal Study of Activity and Ageing was set up in 1983 to assess the role of lifestyle and Customary Physical Activity (CPA) in promoting and maintaining mental health and psychological well-being in later life. The first population survey was conducted between May and September 1985, during which time 1042 people randomly sampled from Family Practitioner Committee lists, and demographically representative of the British elderly population, were interviewed in their own homes. The interview questionnaire contained a total of 318 items and covered aspects of health, lifestyle, demographic and socio-economic status. Using valid and reliable additive scales, or, where appropriate activity inventories, the interview contained specific assessments of CPA, psychological wellbeing, social engagement and physical health (see Morgan *et al.*, 1987; Dallosso *et al.*, 1988).

Customary physical activity was defined as those activities with a probable minimum energy cost of 2 kcal/min, performed continuously for a minimum of 3 minutes, at least weekly, for at least the previous 6 weeks. These activities were divided into five mutually exclusive categories: outdoor productive activities, e.g. gardening, house and car maintenance; indoor productive activities, e.g. housework, decorating, indoor maintenance; walking, i.e. purposeful walking outside the house or garden; shopping, i.e. continuous ambulatory behaviour associated with shopping; and leisure activities, e.g. cycling, swimming, callisthenics; walking was included as a leisure activity only if it was described as such.

In administering the questionnaire on outdoor, indoor and leisure activities the interviewer first determined whether the respondent's participation in the activity met the criteria for 'customary', and then asked in detail about the frequency and duration of participation. Each reported activity was scored as minutes per week. Non-participation was scored as zero. In the assessment of walking and shopping, the interviewer asked in detail about walking done on the day prior to the interview. If, however, this day had not been typical then another day was selected (up to a maximum of 6 days previously). Both walking and shopping were recorded as minutes per typical day. Non-participation was again scored as zero.

In addition, non-continuous activities likely to contribute to muscle strength (e.g. climbing high steps, dragging heavy loads) and joint flexibility (e.g. reaching for high shelves, bending for low shelves) were also assessed. Typically, these activities form discrete units of physical activity and so were recorded in terms of frequency of performance on a five-point scale (i.e. performed never, occasionally, once or several times a week, daily, or several times a day). The reliability of the CPA assessment has been described elsewhere (Dallosso *et al.*, 1988). At the end of each interview, anthropometric measurements of weight, handgrip strength, and shoulder flexibility were also made.

Measurements of morale were provided by Wood, Wylie and Sheafor's (1966) 13-item version of Neugarten, Havighurst and Tobin's (1961) life satisfaction index (the LSIZ). To improve the comprehension of this scale in a British sample, the word form of some of these items was modified slightly (see Morgan *et al.*, 1987).

The first complete follow-up of survivors, using identical materials and methods, was conducted between May and September 1989. All respondents who had participated in 1985, and who were still living in

Table 10.1. *Sample details*

Number of respondents interviewed in 1985	1042
Successful interviews in 1989	690
Losses between 1985 and 1989:	
Death	261
Refusal	63
Untraceable	25
Emigration	3

Nottingham, were invited to participate in the follow-up study. Information on respondents who had died since 1985 was provided by the NHS central register, general practitioners' records, and hospital case notes. Information on respondents who had moved within, or migrated out of the study area was also provided by NHS records, the local authority and, where appropriate, local social networks. Follow-up interviews on 690 people were carried out, representing follow-up response rate of 88.3%. Principal causes of attrition included: death; refusal; emigration; and, in a small minority of cases, the respondent was untraceable (see Table 10.1).

Data from the original (1985) survey allowed for detailed cross-sectional analyses of psychological wellbeing (Morgan *et al.*, 1987), CPA (Dallosso *et al.*, 1988) and relationships between activity and morale (Morgan *et al.*, 1991). The longitudinal data provided by the 1989 survey now allows for detailed analyses of temporal trends in all of these variables. In this chapter patterns of stability and change in both activity and morale are described, and the degree to which change in levels of activity predicts change in morale are assessed. Longitudinal trends in the anthropometric measurements of weight, handgrip strength, and shoulder flexibility are to be described elsewhere.

As earlier investigations had shown marked gender differences in both the overall levels, and in the structure of CPA, analyses of the follow-up data were conducted separately for men and women (see Dallosso *et al.*, 1988). Three aspects of the follow-up data will be considered in this chapter: longitudinal patterns of change in levels of CPA; longitudinal patterns of change in the (factor) structure of CPA; and intercorrelations between CPA and wellbeing. As in previous reports from this project, descriptive data will be presented separately for younger (aged 65–74 in 1985) and older (aged 75+ in 1985) age-groups. To simplify longitudinal comparisons, intervals defining blocks of time in each of the activity distributions are the same as those used by Dallosso *et al.* (1988).

Longitudinal change in levels of CPA

Descriptive data from 1985 and 1989 showing the range, mean, skew, and percentage of individuals scoring zero for walking, shopping, indoor, outdoor, and leisure activities are shown in Tables 10.2 and 10.3 (for women and men respectively). Longitudinal change, i.e. 1985 level minus 1989 level for each block of time in the distributions of these activities is shown in Figures 10.1 to 10.5 (for women) and Figures 10.6 to 10.10 (for men).

Most activities showed a marked negative 'drift' in the distribution characterized by a decrease in higher levels of participation, with a reciprocal increase in zero participation (see Figures 10.1 to 10.10). Nevertheless, it is possible to identify at least two overall patterns. For women, outdoor and leisure activities show a marked increase in zero participation, particularly among the older respondents (Figures 10.4 and 10.5). For walking and shopping activities, however, the distribution of scores actually shows a decrease in zero participation, with a reciprocal increase in the middle blocks (i.e. 0.5–1.5 hours of participation per day; see Figures 10.1 and 10.2). For each activity, it is clear that reductions in participation tended to be greater for the older respondents, though again, this was activity specific. Compared with levels reported in 1985, for example, zero participation in shopping decreased by 20% among older respondents.

For men, too, there were decreases in the mean time spent on leisure and outdoor productive activities for both older and younger age groups (Table 10.3) with reciprocal increases in zero participation (Figures 10.9 and 10.10). Once again, however, walking and shopping showed a quite

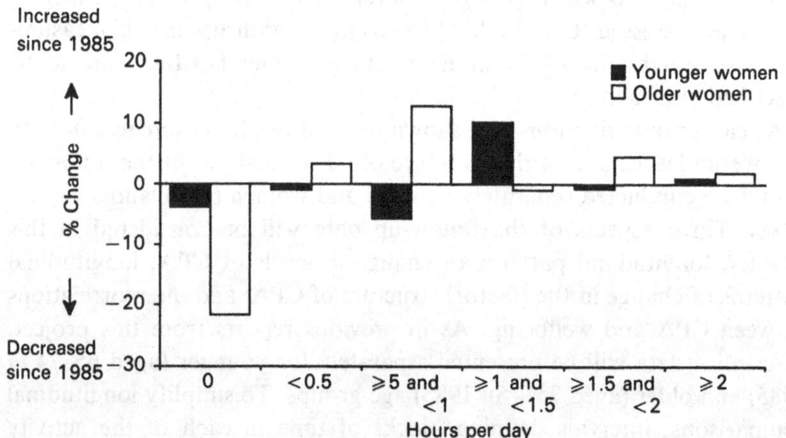

Figure 10.1. Longitudinal changes in the distribution of reported walking/day among women.

different pattern of change, with overall increases in the mean levels (Table 10.3), and, as with the women (Table 10.2), increased participation in the middle bands (Figures 10.6 and 10.7). Changes in the pattern of zero scores was less marked among the men. For younger men, there were increases in zero scores for indoor and outdoor productive activities and leisure. For older men, there were increases in the percentage of zero scores for walking, outdoor productive activities and leisure. Interestingly, there was the marked *decrease* in zero scores for shopping activities among both older and younger men. Compared with levels reported in 1985, zero participation in shopping decreased for both age-groups by about 12%.

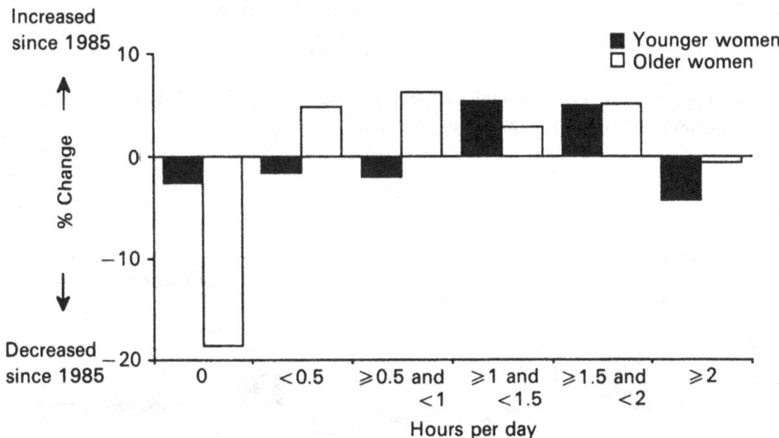

Figure 10.2. Longitudinal changes in the distribution of reported shopping/ day among women.

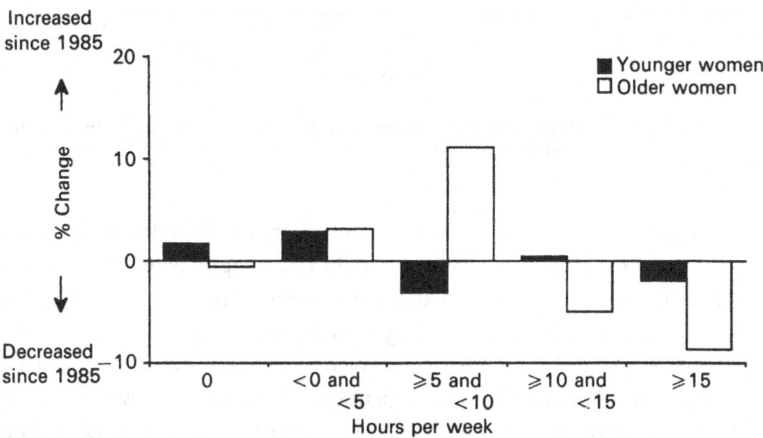

Figure 10.3. Longitudinal changes in the distribution of reported indoor productive activity/week among women.

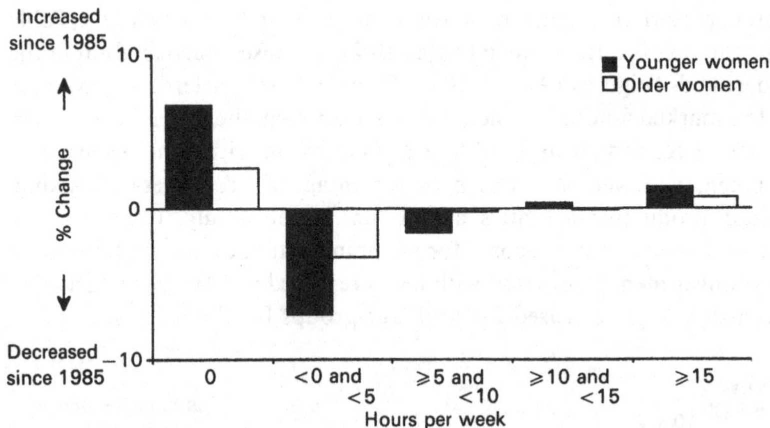

Figure 10.4. Longitudinal changes in the distribution of reported outdoor productive activity/week among women.

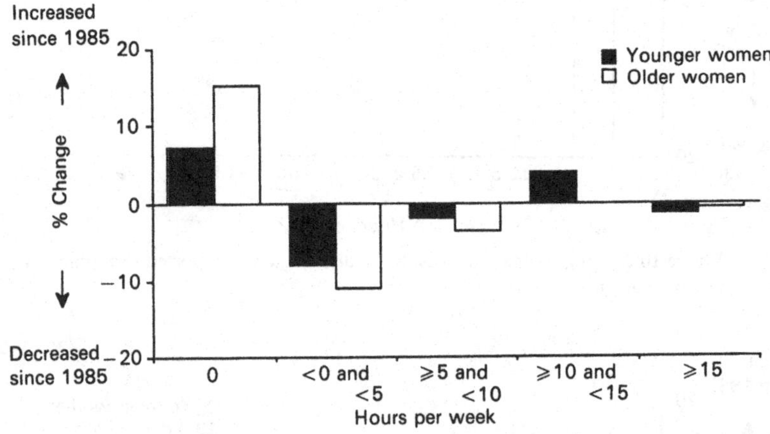

Figure 10.5. Longitudinal changes in the distribution of reported leisure activity/week among women.

Activities requiring flexibility and strength were measured as scale scores with a range zero (non-participation) to 25 (very high levels of participation). Descriptive data from 1985 and 1989 showing the range, mean, skew, and percentage of individuals scoring zero for the strength and flexibility scales are shown in Table 10.4 for both women and men. In most instances, there were small increases in the percentage of zero scores, while, for all sub-groups, there were decreases in the means between 1985 and 1989. This change in mean scores is clearly shown in Figure 10.11.

Table 10.2. *Levels of participation in selected activity categories for women (walking and shopping min/day; remainder min/week)*

Activity	Year	Age group	n*	Range	Mean	Skew	% Zero scores
Walking	1985	Younger	238	0–195	28.3	1.9	36.1
		Older	319	0–245	15.2	3.5	57.7
	1989	Younger	205	0–315	33.5	14	32.2
		Older	158	0–200	19.3	12.5	43.7
Shopping	1985	Younger	282	0–310	29.4	2.3	53.2
		Older	217	0–180	11.2	3	76.9
	1989	Younger	205	0–240	31.4	1.7	50.7
		Older	154	0–330	22.4	3.6	58.4
Indoor	1985	Younger	282	0–2100	606.8	1.0	4.3
		Older	337	0–1958	481.2	0.92	8.6
	1989	Younger	201	0–1900	574	0.85	6
		Older	148	0–1680	402.9	1.4	8.1
Outdoor	1985	Younger	282	0–1680	141.8	2.6	45.7
		Older	337	0–1680	65	4.6	70.9
	1989	Younger	202	0–2400	149.7	3.7	52.5
		Older	148	0–1680	70.4	1.4	73.6
Leisure	1985	Younger	282	0–1740	105.8	3.6	55
		Older	337	0–1500	57.3	5.6	62.9
	1989	Younger	77	0–870	94.3	2.5	62.3
		Older	55	0–140	14.9	2.3	78.2

*N of complete data seta for each activity.

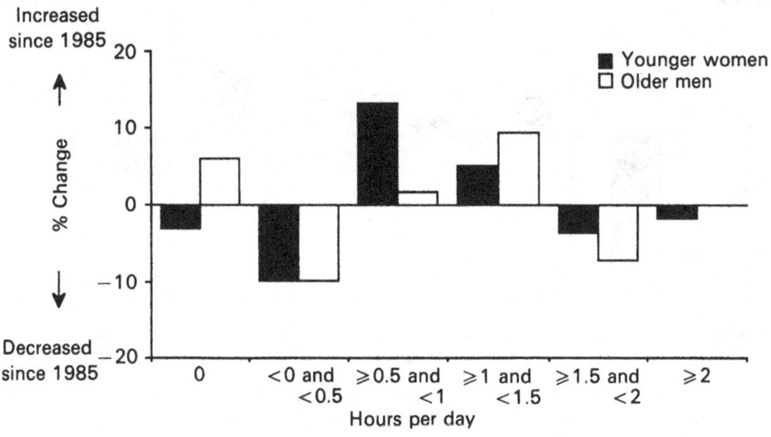

Figure 10.6. Longitudinal changes in the distribution of reported walking/day among men.

Table 10.3. *Levels of participation in selected activity categories for men (walking and shopping min/day; remainder min/week)*

Activity	Year	Age group	*n**	Range	Mean	Skew	% Zero scores
Walking	1985	Younger	206	0–250	37.1	1.8	28.6
		Older	164	0–200	24.7	1.9	43.3
	1989	Younger	161	0–220	40.2	1.7	24.8
		Older	84	0–150	21.6	1.7	48.8
Shopping	1985	Younger	219	0–240	20.3	2.8	67.1
		Older	180	0–240	14.9	3.3	71.7
	1989	Younger	160	0–210	26.9	2	54.4
		Older	84	0–180	22.9	1.7	59.5
Indoor	1985	Younger	217	0–3240	290	3.6	23.5
		Older	180	0–1500	14.9	3.3	30
	1989	Younger	160	0–1500	255.8	1.7	25.6
		Older	81	0–1500	221.8	1.8	28.4
Outdoor	1985	Younger	219	0–2850	446.6	1.8	24.7
		Older	180	0–2640	260	3	47.2
	1989	Younger	160	0–2160	379.4	2	25.6
		Older	80	0–1701	202.3	2.2	49.4
Leisure	1985	Younger	218	0–1710	215.9	1.9	40
		Older	180	0–1560	167.3	2.2	52
	1989	Younger	167	0–1290	144.8	2.7	55.4
		Older	88	0–900	99.2	3	62.5

*N of complete data sets for each activity.

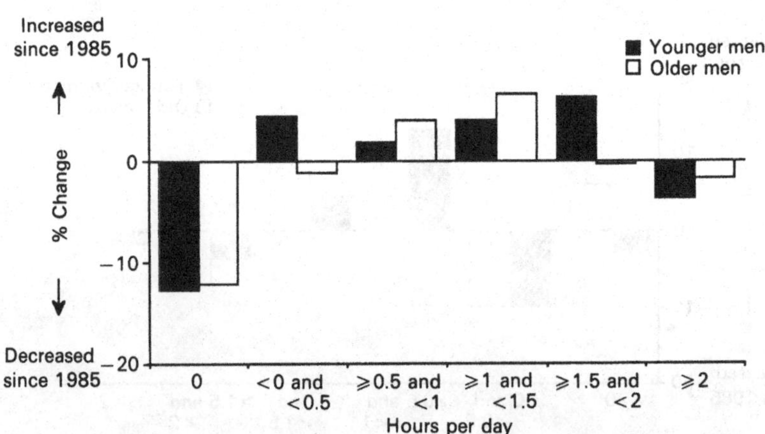

Figure 10.7. Longitudinal changes in the distribution of reported shopping/day among men.

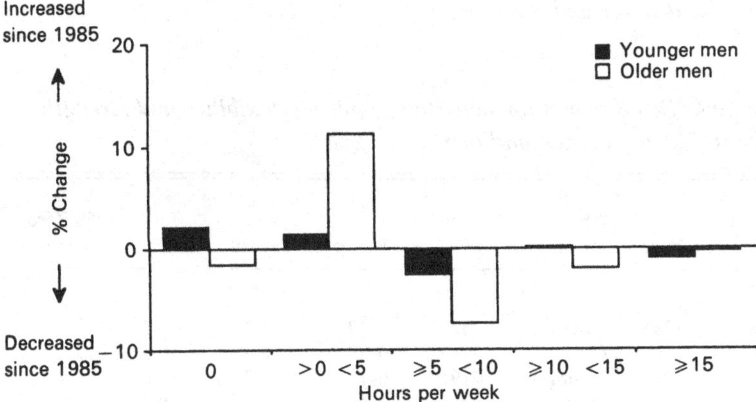

Figure 10.8. Longitudinal changes in the distribution of reported indoor productive activity/week among men.

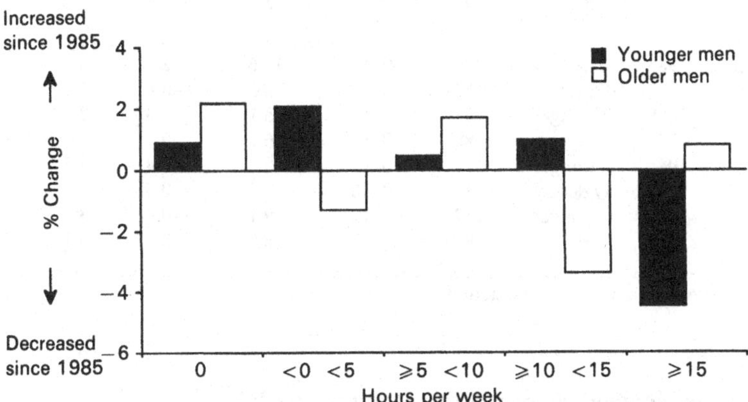

Figure 10.9. Longitudinal changes in the distribution of reported outdoor productive activity/week among men.

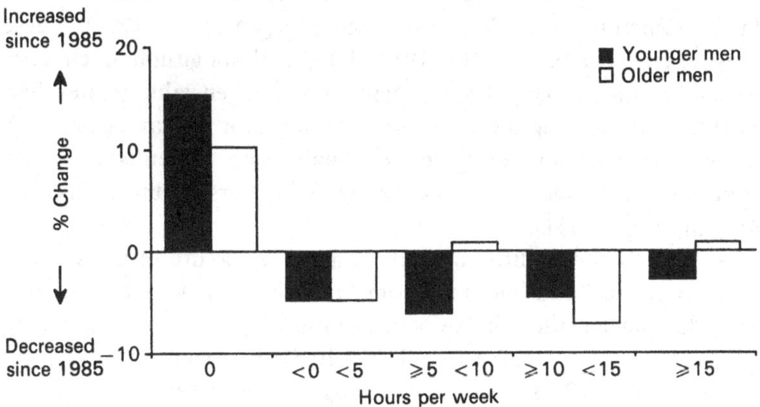

Figure 10.10. Longitudinal changes in the distribution of reported leisure activity/week among men.

Table 10.4. *Scale scores for activities requiring flexibility and strength (range 0–25) for women and men*

Activity	Year	Age group	n*	Range	Mean	Skew	% Zero scores
Women							
Flexibility	1985	Younger	282	0–19	10.3	−0.3	7
		Older	331	0–19	8.4	0.0	3
	1989	Younger	219	0–16	7.7	0.0	1.4
		Older	192	0–16	5.4	0.4	6.3
Strength	1985	Younger	282	0–18	9.9	−0.7	2.1
		Older	334	0–18	7.3	0	7.2
	1989	Younger	218	0–20	8.6	−0.3	3.2
		Older	192	0–14	5.1	0.3	3.2
Men							
Flexibility	1985	Younger	217	0–18	10.6	2	0.9
		Older	178	0–17	8.5	−0.4	2.2
	1989	Younger	167	0–16	8.1	−0.3	2
		Older	88	0–15	6.2	0.1	5.7
Strength	1985	Younger	218	0–19	10.2	−0.4	4.5
		Older	176	0–16	8.5	−0.4	4
	1989	Younger	167	0–19	9.1	−0.8	3
		Older	88	0–17	6.9	0	9.1

*N of complete data sets for each activity.

Longitudinal change in the structure of CPA

Principal components analysis of all seven activities (walking, shopping, indoor, outdoor, leisure, strength, and flexibility) had been used to assess the factor structure of CPA in 1985 (see Morgan *et al.*, 1991). These analyses were repeated for the 1989 data, and longitudinal changes assessed. Only those principal components with an eigen value greater than 1 were extracted, with separate analyses conducted for men and women. A summary of the main findings from these analyses are presented in Tables 10.5 and 10.6. The names given to the extracted factors are those reported by Morgan *et al.*, (1991).

For women there was little overall change in structure of CPA. Both 'home maintenance' and 'pleasure' comprised the same variables as they had in 1985. Indeed, the relative contribution of leisure and walking to 'pleasure' remained the same (see Table 10.5). For men, however, the overall structure of CPA shows an interesting change. In 1985, three factors were extracted: 'home maintenance'; 'pleasure'; and 'home management', indicating that, for men, participation in 'domestic' activities was unrelated

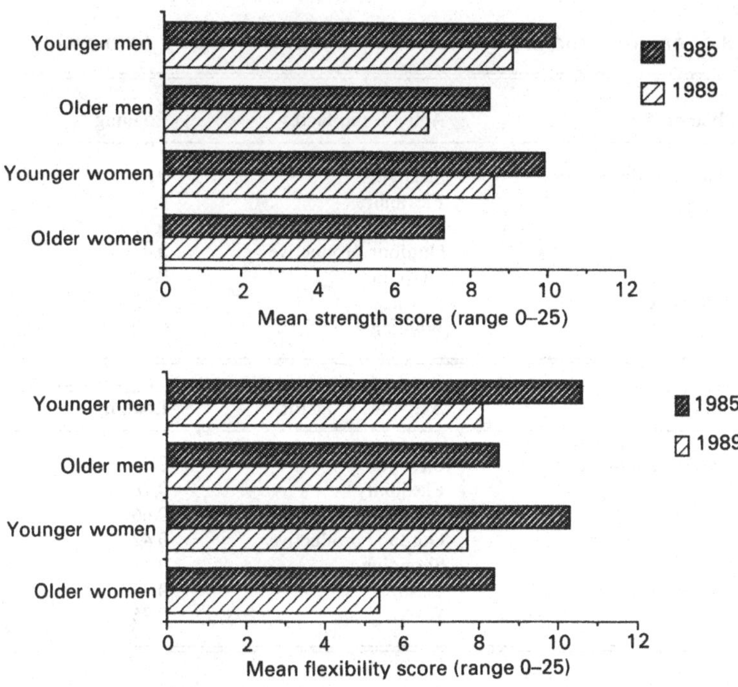

Figure 10.11. Mean scores for men and women for activities requiring strength and flexibility for 1985 and 1989.

to their participation in other domains of activity. Put another way, busy men were no more likely to engage in housework and other domestic activities than were sedentary men. By 1989, factor 'home management' had been subsumed under both 'home maintenance' and 'pleasure', a structure similar to that found for women (see Table 10.6).

Activity and wellbeing

Overall mean levels of morale, as measured by the Life Satisfaction Index, for 1985 and 1989 are shown in Figure 10.12. Despite being active across its range of 13 items, i.e. the scale shows no evidence of a ceiling or floor effect on both occasions of measurement, the LSIZ showed remarkable stability over the four year inter-study period. To assess the strength of any activity–affect relationship a raw change score (1985 minus 1989) for LSIZ and all seven activity categories was computed for each respondent. Tables 10.7 and 10.8 show the pattern of correlations among change scores for women and men respectively. Significant correlations among the activity variables reflect, to some extent, the factor structure already presented.

Table 10.5. *Factor structure of Customary Physical Activity for women*

(*a*) 1985	Factor	Activities in factor	Loading
1.	Home maintenance	Strength	0.81
		Flexibility	0.72
		Indoor productive	0.71
		Outdoor productive	0.47
		Shopping	0.42
2.	'Pleasure'	Leisure	0.82
		Walking	0.80

(*b*) 1989	Factor	Activities in factor	Loading
1.	'Home maintenance'	Strength	0.74
		Flexibility	0.78
		Indoor productive	0.66
		Outdoor productive	0.42
		Shopping	0.57
2.	'Pleasure'	Leisure	0.77
		Walking	0.75

Table 10.6. *Factor structure of Customary Physical Activity for men*

1985	Factor	Activities in factor	Loading
1.	'Home maintenance'	Outdoor productive	0.79
		Flexibility	0.78
		Strength	0.76
2.	'Pleasure'	Leisure	0.83
		Walking	0.82
3.	'Home management'	Shopping	0.73
		Indoor productive	0.66

1989	Factor	Activities in factor	Loading
1.	'Home maintenance'	Outdoor productive	0.76
		Flexibility	0.70
		Strength	0.68
		Indoor productive	0.64
2.	'Pleasure'	Leisure	0.71
		Walking	0.71
		Shopping	0.35

Table 10.7. *Product-moment correlation coefficients among change scores (1985 minus 1989) from seven activity categories and life satisfaction (LSIZ) for women*

	Walking	Shopping	Indoor	Outdoor	Leisure	Strength	Flexibility	LSIZ
Walking		0.0	0.0	0.0	−0.3**	0.1**	0.0	0.0
Shopping			−0.1	−0.1*	−0.1	0.0	−0.1	0.1*
Indoor				−0.0	0.1	0.3**	0.0*	−0.1
Outdoor					0.1	0.1*	0.0	−0.1
Leisure						−0.0	0.0	0.0
Strength							0.3**	0.0
Flexibility								0.1*

*$P<0.05$. **$P<0.01$.
NB Rounded to one decimal place.

Table 10.8. *Product-moment correlation coefficients among change scores (1985 minus 1989) from seven activity categories and life satisfaction (LSIZ) for men*

	Walking	Shopping	Indoor	Outdoor	Leisure	Strength	Flexibility	LSIZ
Walking		−0.1	0.1*	−0.0	−0.0	0.2**	0.0	0.1*
Shopping			−0.1	0.0	−0.1	0.1	−0.0	0.0
Indoor				0.0	−0.3**	0.1	0.1*	−0.2**
Outdoor					−0.1	0.1	0.0	0.1*
Leisure						0.1	0.1	0.3**
Strength							0.3**	0.2*
Flexibility								0.1*

*$P<0.05$. **$P<0.01$.
NB Rounded to one decimal place.

Interestingly, however, a pattern of correlations between activities and morale does emerge which, very similar to the pattern found in the cross-sectional data (Morgan *et al.*, 1991), is weak, selective, and gender specific.

Comment

Overall, longitudinal changes in levels of customary physical activity are characterized more by drift than abrupt 'shift', while longitudinal measurements of morale appear to be characterized by stability rather than by

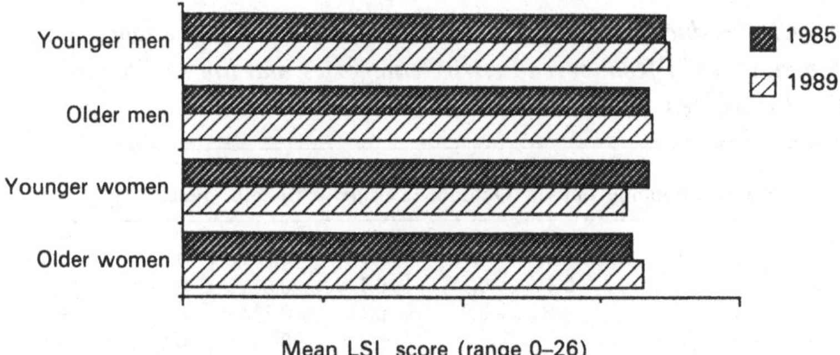

Figure 10.12. Life satisfaction (LSIZ) mean scores for men and women for 1985 and 1989.

change. Nevertheless, for some selected activities, particularly among men, there is clear evidence of an activity–affect relationship. The structure, direction, and additional factors possibly mediating these relationships will provide the focus of attention for future analyses.

References

Dallosso, H.M., Morgan, K., Bassey, E.J., Ebrahim, S.B.J., Fentem, P.H. & Arie, T.H.D. (1988). Levels of customary physical activity among the old and the very old living at home. *Journal of Epidemiology and Community Health*, **42**, 121–7.

Folkins, C.H. & Sime, W.E. (1981). Physical fitness training and mental health. *American Psychologist*, **36**, 373–89.

Hughes, J.R. (1984). Psychological effects of habitual aerobic exercise: a critical review. *Preventive Medicine*, **13**, 66–78.

Morgan, K., Dallosso, H.M., Arie, T., Byrne, E.J., Jones, R. & Waite, J. (1987). Mental health and psychological well-being among the old and the very old living at home. *British Journal of Psychiatry*, **150**, 801–7.

Morgan, K., Dallosso, H., Bassey, E.J., Ebrahim, S., Fentem, P.H., Arie, T. (1991). Customary physical activity, psychological wellbeing, and successful ageing. *Ageing and Society*, **11**, 399—415.

Neugarten, B.L., Havighurst, R.J. & Tobin, S.S. (1961). The measurement of life satisfaction. *Journal of Gerontology*, **16**, 134–43.

Shephard, R.J. & Montelpare, W. (1988). Geriatric benefits of exercise as an adult. *Journal of Gerontology*, **43**, M86–90.

Veale, D.M.W. de C. (1987). Exercise and mental health. *Acta Psychiatrica Scandinavica*, **76**, 113–20.

Wood, V., Wylie, M. & Sheafor, B. (1966). An analysis of a short self-report measure of life satisfaction: correlation with rater judgement. *The Gerontologist*, **6**, 31 (abstract).

11 *The benefits of low intensity exercise*

ADRIANNE E. HARDMAN

Introduction

It is a fundamental biological principle that man is adaptable. Regular physical activity provokes adaptations, many of which are beneficial, but the amount and intensity of exercise needed to confer benefit remains uncertain. The purpose of this chapter is to examine the potential of modest amounts of physical activity to promote health by 1) maintaining or increasing functional capacity, 2) influencing energy balance, 3) decreasing the likelihood of some diseases and 4) contributing to the management of patients with existing disease.

Functional capacity

The single most important aspect of functional capacity is, arguably, the maximal oxygen uptake ($\dot{V}O_2max$). This important quantity imposes a limit on the intensity of whole body dynamic exercise which can be sustained. Other aspects such as muscle strength and mobility become increasingly important in old age as the decline in capacities means that thresholds which limit daily activities are approached.

Classic studies in Scandinavia at the end of the 1960s showed that endurance training increases, and enforced inactivity decreases, $\dot{V}O_2max$ (Saltin *et al.*, 1968). The relative improvement with training depends on two factors: 1) the initial physical activity level, the more sedentary the individual the greater the scope for improvement, and 2) heredity. Exceptionally, training has been reported to increase $\dot{V}O_2max$ by as much as 44% (Hickson, Bomze & Holloszy, 1977) but increases in the range 10–20% are more usual. Differences of 100% or more between individuals of the same age and sex are therefore of too great a magnitude to be attributable to training status alone and must reflect genetic predisposition.

Maximal oxygen uptake declines by about 10% per decade after the age of 25 years (Hagberg, 1987), in line with the age-related loss of muscle mass (Fleg & Lakatta, 1988). Whilst some of this decline is attributable to the ageing process *per se*, a decrease in daily physical activity probably contributes because older men who maintain their activity levels experience a rate of decrease in $\dot{V}O_2$max which is about half that found in sedentary individuals, i.e. about 5% per decade (Hagberg, 1987).

Although the decline in functional capacity with ageing may be linear the associated fall in quality of life is not. For the elderly, activities inside and outside the home are limited by thresholds of functional capacity. For example, if $\dot{V}O_2$max falls to the point where the comfortable speed of walking is so slow that an old person feels unsafe out in the street, they will become increasingly housebound. The timing of traffic signals at pedestrian crossings assumes a walking speed of some 1.3 to 1.4 $m\,s^{-1}$, about 20 minutes per mile, but for a healthy woman of about 80 years of age the speed of comfortable walking is about 0.9 $m\,s^{-1}$, or 30 minutes per mile (Danneskiold-Samsøe *et al.*, 1984). Elderly women therefore feel anxious because they have insufficient time to cross in safety and are at increased risk of accident. Even a small improvement in $\dot{V}O_2$max will allow a faster speed of walking to be sustained, increasing confidence and independence.

Studies have shown that increased physical activity can improve $\dot{V}O_2$max in older individuals by amounts which are important for their quality of life. A 12-month endurance training programme (6 months' low intensity followed by 6 months' high intensity) resulted in a 30% improvement in $\dot{V}O_2$max in men and women of average age 63 years (Seals *et al.*, 1984). Even older individuals (aged 70 to 79 years) showed a 22% increase in $\dot{V}O_2$max after 26 weeks of training (Hagberg *et al.*, 1989). The potential to adapt therefore persists into old age.

Even in the absence of increases in $\dot{V}O_2$max, endurance, the ability to sustain exercise without fatigue, can be enhanced by increased physical activity levels in young and old alike. Endurance derives from local metabolic adaptations of skeletal muscle (Gollnick *et al.*, 1973) which improve its oxidative capacity. They are of a greater magnitude than increases in $\dot{V}O_2$max (Gollnick *et al.*, 1973) but more labile (Henriksson & Reitman, 1977). Muscle's capacity for oxidative metabolism is further enhanced by the development of the microcirculation (Ingjer, 1979), an adaptation which may be coupled closely with the modifications of lipoprotein metabolism referred to below. Moreover, improvements are as great in elderly men and women as in young adults (Meredith *et al.*, 1989).

For many individuals, physical activity need not be vigorous and sporting to provoke adaptation. For example, fast walking is a sufficient aerobic stimulus to improve functional capacity in most adults. When asked to walk 1 mile as fast as possible, about 67% of men and 91% of women over 30 attain a heart rate which would be expected to elicit a training effect (Porcari *et al.*, 1987). This is equivalent to approximately

Table 11.1 *Responses to a one mile walk at a self-selected brisk pace in 52 sedentary women*

	Mean	SD
Speed ($m\,s^{-1}$)	1.7	0.2
$\dot{V}O_2$ ($ml\,kg^{-1}\,min^{-1}$)	16.2	3.2
$\%\dot{V}O_2max$	59	14
Heart rate (beat min^{-1})	131	18
Blood lactate concentration ($mmol\,l^{-1}$)	2.08	1.06

60% of $\dot{V}O_2max$ (Hudson & Hardman, 1989). Table 11.1 shows some physiological responses to walking at a self-selected 'brisk pace', about 15 min 47 s per mile, in 52 sedentary women aged 44.2 ± 8.6 years (Mean $\pm$ SD).

Walking at 4.5 mph requires an oxygen uptake of about 23 ml $kg^{-1}\,min^{-1}$, eliciting 60% of $\dot{V}O_2max$ for anyone possessing a value of 38 ml $kg^{-}\,min^{-1}$ or less. Survey data to describe the $\dot{V}O_2max$ values of the British population are not yet available but the findings of the Canada Fitness Survey suggest that 75% of women aged 20–29 and more than 90% of women over 30 possess values below this threshold. Similarly, amongst Canadian men, 25% of 30–39 year-olds, 55% of 40–49 year-olds and more than 85% of over 50s fall into this category (Shephard, 1986). There is no reason to suspect that British men and women possess higher $\dot{V}O_2max$ values than their Canadian counterparts. Indeed, data from the author's laboratory suggest that few women over 30 years of age have values in excess of 38 ml $kg^{-1}\,min^{-1}$ (Hardman *et al.*, 1988). The author's experience with sedentary male volunteers aged between 42 and 59 years is that values range from 25.8 to 46.8 ml $kg^{-1}\,min^{-1}$ (unpublished observations).

As might be expected from this examination of the relative oxygen cost of walking, longitudinal studies suggest that regular, brisk walking can improve $\dot{V}O_2max$ by between 13 and 28% (Pollock *et al.*, 1971; Jetté, Sidney & Campbell, 1988; Santiago *et al.*, 1987). At Loughborough 28 women (aged 44.0 ± 7.9 years) who followed a progressive programme of brisk walking for 1 year were monitored. They walked for an average of 155 ± 48 minutes (range 75–287) per week, equivalent to between 16 and 17 km. Maximal oxygen uptake increased for walkers relative to controls

($P<0.05$) (walkers 26.7 ± 3.7 vs 29.7 ± 4.2 ml kg^{-1} min^{-1}; controls 27.6 ± 4.0 vs 27.7 ± 4.8 ml kg^{-1} min^{-1}) and the increase was accompanied by favourable changes in the cardiovascular and metabolic response to standard, submaximal exercise (Hardman *et al.*, 1988).

Brisk walking is more likely to elicit a training response in women than in men because it constitutes relatively higher intensity exercise for women. Nevertheless, in men aged 50.9 ± 5.3 years (range 42–59) who completed an average of 28 ± 9 minutes (range 22–47) of brisk walking per week for 1 year, the response of $\dot{V}O_2$max differed ($P<0.05$) from that of the controls (Stensel *et al.*, 199?). For walkers, $\dot{V}O_2$max was maintained whilst values for the controls fell (walkers 35.9 ± 4.6 vs 35.0 ± 4.9 ml kg^{-1} min^{-1}, controls 35.2 ± 5.2 vs 32.2 ± 4.6 ml kg^{-1} min^{-1}). In addition, there was evidence of marked changes in skeletal muscle metabolism in the walkers. Blood lactate concentration during treadmill walking at 80% of baseline $\dot{V}O_2$max was lower after 1 year for walkers ($P<0.05$) but not for controls (walkers 4.9 ± 2.2 vs 3.1 ± 1.6 mmol l^{-1}, controls 4.6 ± 2.3 vs 4.0 ± 1.8 mmol l^{-1}). Attainable amounts of a socially acceptable form of exercise can therefore be expected to confer benefits in terms of increased functional capacity and endurance for many adults.

Energy balance

If functional capacities are to be extended, or maintained, the intensity of physical activity is important. In contrast, any physical activity, whatever the intensity, contributes to total energy expenditure. Moreover, although the cost of habitual physical activity rarely accounts for more than 20% of total energy expenditure (Stock & Rothwell, 1982), exercise is the one way in which energy expenditure can be increased voluntarily. A detailed consideration of the extensive literature concerning the influence of exercise on body fat stores is inappropriate here. The intention is to consider the potential of modest amounts of physical activity to increase energy turnover and contribute to the maintenance of desirable body weight.

Some 83% of the mass of adipose tissue is triglyceride, with an energy density of 37 kJ g^{-1}. The energy stored in 1 kg of this tissue is therefore approximately 30.7 MJ. Energy expenditure during weight-bearing activities depends on body mass so that walking or running 1 mile expends some 293 kJ for a 50 kg person but about 460 kJ for an 80 kg person. Consequently, to expend the energy equivalent of 1 kg of adipose tissue means walking or running between 70 and 110 miles. This approach makes exercise look a most unattractive strategy for weight loss but these calculations undervalue the potential of physical activity to contribute to long-term maintenance of energy balance.

An alternative way of presenting the same relationships makes the point. If walking 1 mile expends between 293 and 460 kJ, walking 1 mile every day for a year expends a total of 107 to 168 MJ, or the energy equivalent of 3.5 to 5.5 kg of adipose tissue. These calculations are, of course, also an oversimplification because they are based on gross, not net, energy expenditure. In addition they do not take account of 1) the diminution of resting metabolic rate as body mass falls, 2) the post-exercise elevation of metabolic rate or 3) any increase in energy intake which may be precipitated by the increased exercise.

Although there is no doubt that metabolic rate remains elevated for a period after exercise, its importance in terms of overall energy balance is the subject of debate. Eighty minutes of exercise at 70% $\dot{V}O_2max$ caused oxygen uptake to be 14% higher during the 12 hours after exercise than during an equivalent control period (Maehlum *et al.*, 1986). Taking the energy equivalent of a litre of oxygen as 20 kJ, this indicates an extra post exercise energy expenditure of about 520 kJ, or about 15% of the net energy expenditure during the bout of exercise. The relevance of this for individuals who may be prepared to invest in only modest amounts of less intense activity is debatable.

Some studies have shown that a moderate increase in physical activity is not inevitably accompanied by an equivalent increase in energy intake. Middle-aged men who averaged 8–12 miles of running per week over 2 years experience a decrease of 3.6% in body fat, despite a 15% increase (1500 kJ) in daily energy intake (Wood, Terry & Haskell, 1985). In a much shorter study, 57 days of moderate physical activity did not alter mean daily energy intake in obese women. Consequently, negative energy balance was achieved and body mass decreased (Woo, Garrow & Pi-Sunyer, 1982). Fat loss does, however, not always accompany increased physical activity levels when restrictions on energy intake are not imposed, e.g. Stubbe *et al.*, 1983. This may be because there is a threshold of exercise, relating to amount and/or intensity, which is needed to provoke fat loss.

There is evidence of other important roles for physical activity in relation to energy storage and turnover. For example, the addition of exercise to a low energy diet has been reported to enhance weight and fat loss and prevent a fall in resting metabolic rate (Pavlou *et al.*, 1989). Moreover, exercise may help with the most difficult aspect of treating obesity and overweight, i.e. weight maintenance after weight loss. Individuals who exercise three times a week, expending about 6300 kJ, tend to be those who are successful in maintaining a reduced body weight (Pavlou, Krey & Steffee, 1989).

Recently, interest in the influence of exercise on body fatness has extended to the distribution of body fat. Amongst 1366 women and 1257 men who took part in the 1981 Canada Fitness Survey, individuals practising vigorous activities on a regular basis had lower waist-to-hip ratios (WHR) than others, even after the effect of subcutaneous fat was adjusted for (Tremblay *et al.*, 1990). Six months' physical training has been found to decrease WHR even in the absence of a reduction in body weight (Krotkiewski, 1987). These latter findings have importance for the role of exercise in disease prevention because abdominal obesity, as reflected by a high waist-to-hip ratio, is known to be strongly associated with an increased risk of myocardial infarction (Lapidus *et al.*, 1984).

It is clear that even modest amounts of physical exercise can, if undertaken regularly, make an important contribution to the maintenance of a desirable level of body fatness. Nevertheless, the role of physical activity in stimulating a loss of body fat remains to be defined more precisely.

Decreased risk of disease

Coronary heart disease

Evidence that regular physical activity decreases the risk of coronary heart disease (CHD) is increasingly convincing. More than 40 epidemiological studies, predominantly cohort studies, have compared the risk of CHD in physically active men with that exhibited by their sedentary counterparts. Early studies compared groups differing with regard to the level of occupational exercise (Morris *et al.*, 1953; Paffenbarger & Hale, 1975) but subsequently emphasis has been on leisure-time physical activity. Paffenbarger and colleagues examined the influence of leisure-time physical activity on the incidence of CHD in cohorts of ex-students who entered the University between 1916 and 1950. Six to ten year follow-up of Harvard graduates showed that the risk of first heart attack was 64% higher in men who expended fewer than 2000 kcal per week in physical activity than in classmates with a lower physical activity index (Paffenbarger, Wing & Hyde, 1978).

Studies of English civil servants found a lower risk of CHD to be associated with vigorous exercise rather than with an amount of energy expenditure in exercise (Morris *et al.*, 1973, 1980). Men who reported vigorous exercise, i.e. swimming, jogging, cycling, brisk walking, racket games, etc, on initial survey in 1968–70 had an incidence of CHD over the next $8\frac{1}{2}$ years which was somewhat less than half that of their colleagues who reported no such exercise. The different emphasis in findings of United States and British studies persists in later papers. Amongst Harvard alumni, there was a 21% lower risk of death (all causes) as distance walked increased from fewer than 3 miles to 9 or more miles per week (Paffenbarger *et al.*, 1986) but a second study of English civil servants could find no association between the quantity of walking done and coronary attack rate (Morris *et al.*, 1990). A particularly low rate of attack was, however, found in the men claiming to be fast walkers (over 4 mph).

Thus, although the amount and intensity of activity needed to confer a reduced risk of CHD is still to be defined, 40 years of research in many countries strongly supports the suggestion that regular physical activity reduces the risk of CHD in men. The disease is far less common in women and too little is known to permit any such conclusions for women. Overall, the epidemiological evidence is that physical inactivity increases the risk of a heart attack in men by a factor of about 1.9. However, the better-designed studies suggest that the relative risk is closer to 2.4, similar to that for smoking (>20 cigarettes a day), hypertension (systolic blood pressure >150 versus <130 mm mercury) and hypercholesterolaemia (plasma cholesterol >6.9 versus <5.6 mmol l^{-1}) (Powell *et al.*, 1987).

These estimates of the potency of physical inactivity as a risk factor for CHD have profound implications. The potential of a particular factor to influence the incidence of CHD depends not only on the relative increase in risk it confers but also on its prevalence. Gathering information about exercise habits is difficult but data from Heartbeat Wales suggests that inactivity is widespread. In the 35–65 year age-group, 33–57% of men and 73–86% of women are 'minimally active or sedentary' (Welsh Heart Programme Directorate, 1987) and therefore at risk of CHD because of their inactivity. If these data are representative, an increase in the number of individuals taking regular exercise could have an important impact on the incidence of CHD in this country. There are, however, two weaknesses in the argument presented here. One is that there is, and probably never will be, any experimental evidence in the form of a randomized controlled trial that taking more exercise decreases CHD mortality. Secondly, some constitutional factor may be responsible for the lower incidence of the disease in physically active men. The view that exercise does exert a beneficial influence on the risk of CHD would be more convincing if a plausible mechanism could be demonstrated.

Speculation about such a mechanism has often involved an effect on lipoprotein metabolism. Comparison of endurance-trained individuals with their sedentary counterparts almost invariably reveals more favourable lipoprotein profiles in the exercisers. Although differences in total cholesterol concentration are small, and not always statistically significant, plasma triglyceride concentration is usually somewhat lower and plasma high density lipoprotein (HDL) cholesterol concentration markedly higher in endurance athletes (Williams *et al.*, 1986; Thompson *et al.*, 1983; Hartung *et al.*, 1980). The results from longitudinal studies are less consistent, but many show either a decrease in triglyceride and an increase in HDL cholesterol concentration or both, e.g. Huttunen *et al.*, 1979; Wood *et al.*, 1988; Marti *et al.*, 1990. High levels of HDL are inversely associated with CHD risk and it seems likely that this relates to the role of these lipoproteins in reverse cholesterol transport (Miller & Miller, 1975).

Changes in lipoprotein metabolism due to increased physical activity may be secondary to exercise induced weight loss (Wood *et al.*, 1988), but adaptations in skeletal muscle are also important. Lipoprotein lipase, the enzyme responsible for the hydrolysis of triglyceride in lipoproteins, is situated on the luminal surface of the capillary endothelium. Increased capillarization in skeletal muscle due to training accelerates the removal of triglyceride-rich lipoproteins from the circulation, enhancing the synthesis of HDL (Kiens & Lithell, 1989).

Most training studies of lipoprotein metabolism have employed moderate to high intensity exercise, usually running, e.g. Wood *et al.*, 1983; Farrell & Barboriak, 1980, but there is evidence of a favourable effect of chronic low intensity, long duration physical activity. For example, a study of active postal carriers found a relationship between reported number of miles walked per day and HDL_2 cholesterol concentration, i.e. that fraction of HDL cholesterol which distinguishes exercisers from controls (Cook *et al.*, 1986).

Consequently, recognizing that high intensity exercise programmes are associated with low adherence and a high incidence of injuries (Martin & Dubbert, 1982), some researchers have begun to examine the efficacy of low intensity and self-monitored programmes as an intervention to modify lipoprotein metabolism. Less than 2 hours of jogging per week over 4 months has been found to increase HDL cholesterol and decrease VLDL cholesterol in previously sedentary men (Marti *et al.*, 1990). In women, a modest programme of brisk walking provoked a 6% decrease in total cholesterol (ns with respect to controls) and a 27% increase in HDL cholesterol ($P < 0.05$) in a group of previously sedentary women (Hardman *et al.*, 1989).

Physical activity may decrease the risk of CHD by modifying risk factors other than those associated with lipoprotein metabolism, e.g. hypertension, and/or by mechanisms involving the acute phases of the disease. Epidemiological evidence that exercise has to be continuing and current in order to be protective (Paffenbarger, Wing & Hyde, 1978; Morris *et al.*,

1990) has led to speculation that it may have an influence on, for example, the clotting processes or platelet aggregation (Morris *et al.*, 1990).

Osteoporosis

Osteoporosis, a decrease in bone mass and strength leading to an increase in fractures, is a major public health problem in this country and elsewhere. Current treatment, which may, or may not, include therapeutic exercise, does not restore lost bone significantly, it only decreases the rate of loss. Consequently, prevention is the strategy of choice and is concerned with increasing peak bone mass at skeletal maturity and/or reducing the rate of bone loss after the menopause.

The rate of bone remodelling is subject to several influences, including mechanical loading. Consequently, bed rest or weightlessness result in bone loss (Krølner & Toft, 1983; Stupakov *et al.*, 1984) whereas increased mechanical loading increases bone mineral content (Rubin & Lanyon, 1984). Early observations showed that bone mass in some skeletal sites is markedly higher in athletes than in comparable controls (Dalen & Olsson, 1974; Lane *et al.*, 1986) and evidence from professional tennis players shows that such differences cannot be attributed to constitutional factors alone. Bone density in the radius of tennis players is not only markedly higher than in controls but also higher in the dominant than in the non-dominant arm (Pirnay *et al.*, 1987). A long-term commitment to sport therefore appears to be associated with a high bone density and the adaptations seem to be a local response to local loading.

What is less certain is whether a modest amount of less intense physical activity is likely to be effective in the prevention of osteoporosis. Some evidence of a dose–response relationship between levels of physical activity and indices of bone condition exists. In a group of 83 women aged between 30 and 85, those with high levels of activity (evaluated by questionnaire) had higher values for bone mineral content, corrected for skeletal size, than did the moderate and low activity groups (Zylstra *et al.*, 1989). The most effective exercise regimen for stimulating an osteogenic response probably involves high loading during short periods of weight bearing activity (Lanyon, 1990) but experimental evidence does not always confirm this. For example, performing low-intensity strength training, in addition to aerobic exercise (walking, jogging, dance routines), did not provoke an increase in bone mass greater than that demonstrated by a comparable group of women who undertook aerobic exercise only (Chow, Harrison & Notarius, 1987). In contrast, women with a history of osteoporotic fractures who performed simple resistance exercises, using their own or a partner's body weight, achieved a 3.8% increase in bone mineral density.

This compared with a 1.9% loss in a matched control group over the same 5-month period (Simkin, Ayalon & Leichter, 1987). Consequently, there may be some scope for therapeutic exercise to thicken and stabilize the remaining trabeculae, even in established osteoporosis.

From the public health point of view, it is important to examine the efficacy of moderate amounts of physical activity. Whilst it may be desirable to design a programme likely to be maximally effective at a number of skeletal sites, this may be difficult to implement on any scale. The most easily accessible form of weight-bearing exercise is walking, and this activity is widely recommended as both a therapeutic and a preventive measure. Although the number of hours of walking per day has been found to be correlated with lumbar and femoral neck densities (Zylstra *et al.*, 1989) prospective studies have, so far, yielded conflicting results. White and colleagues studied the influence of walking (progressing to 2 miles on 4 days per week) on the bones of recently menopausal women. Over 6 months, walkers lost bone mineral content at the same rate as did controls (White *et al.*, 1984). In one of the few randomized trials reported, postmenopausal women walking about 7 miles per week over 3 years exhibited a loss of cortical bone density which did not differ from that shown by controls (Cauley *et al.*, 1986). Both these studies relied on measurements of the radius, and may therefore have failed to reveal adaptive changes due to local loading.

More recently, broadband ultrasonic attenuation (BUA) has been used to study the effects of brisk walking on bone condition. In women aged 30 to 61 years (mean 44) who walked briskly for 16 to 18 km per week, BUA measured in the calcaneus was increased after 1 year by 12% whereas it decreased by 4% for the controls (Jones *et al.*, 1991). The calcaneus may be an appropriate skeletal site for examining the effects of this mode of exercise because it is the site of ground reaction forces at heel strike during walking. The site of measurement, the lumbar.spine, may be one reason why a recent study employing a similar brisk walking regimen failed to find a beneficial effect on bone density (Cavanagh & Cann, 1988). However, BUA of the calcaneus is reported to be a very good predictor of lumbar vertebral bone density (Baran *et al.*, 1988) so it seems unlikely that the measurement site alone can account for the inconsistency between these two reports. Further work clearly is needed to determine the amount and type of physical activity needed to influence skeletal health.

Management of patients with existing disease

Physical activity contributes to the management of patients with existing disease because it maximizes residual functional capacities and increases independence, for example, by enabling an earlier return to work. An acute event such as myocardial infarction can impair functional capacity as can chronic diseases such as claudication and chronic airway limitation.

The spectrum of patients with CHD, for whom exercise rehabilitation is recommended, has been progressively extended over the last two decades. Current intensive interventions during the acute phase of myocardial infarction, e.g. thrombolytic therapy, have reduced or prevented myocardial damage so that patients are discharged earlier, often with little or no restriction of activity (Wenger, 1989). Exercise rehabilitation programmes now include elderly patients as well as some with stable chronic heart failure (Coats *et al.*, 1990) and are increasingly unsupervised or partially supervised. The mechanisms underlying the improved functional capacity in coronary heart disease patients are still being clarified but the most important is likely to be the decreased myocardial oxygen demand at any sub-maximal exercise level.

Low intensity exercise (not greater than 50% $\dot{V}O_2max$) has been shown to elicit a training effect in patients recovering from myocardial infarction (Rechnitzer *et al.*, 1983), probably because of peripheral adaptations rather than because of any improvement in ventricular function. An obvious advantage is that this level of exercise can be prescribed widely with safety, decreasing the need for supervision of low risk patients and hence keeping costs low.

The question as to whether exercise rehabilitation benefits patients in terms of decreased mortality and/or morbidity has not been answered definitively. However, meta-analysis of data from 36 trials revealed that cardiovascular mortality and fatal reinfarction at 3 years were significantly lower in rehabilitation patients than in control patients (O'Connor *et al.*, 1989). Such analyses have their limitations, and conclusions must be appropriately cautious. In any case, exercise rehabilitation can be justified for many, perhaps even for a majority, of patients with CHD without the need to argue for secondary prevention.

Exercise is one of the commonly used non-pharmacological therapies for hypertension. In adults with mild to moderate hypertension, increased physical activity reduces systolic and diastolic arterial blood pressure by, on average, 13 and 10 mm mercury respectively (Cade *et al.*, 1984; Duncan *et al.*, 1985; Nelson *et al.*, 1986). Low intensity training may be as effective or more effective in this regard than moderate intensity training. Hagberg and colleagues recently compared the effect of training at 53% or 73% of $\dot{V}O_2max$ in hypertensive men and women of average age 64 years (Hagberg *et al.*, 1989). Both regimens were similarly effective in reducing diastolic blood pressure (11 to 12 mm mercury) but, for reasons which remain to be explored, the decrease in systolic blood pressure was significantly greater for the low intensity group (20 vs 8 mm mercury).

Acute exercise can induce bronchoconstriction in asthmatics but physical activity is nevertheless an important component in the clinical management of asthma. A recent controlled trial demonstrated that exercise training, three times a week for 30 minutes on each occasion at 75% of predicted maximal heart rate (about 65% of $\dot{V}O_2$max), resulted in a 22% increase in $\dot{V}O_2$max and a decrease in ventilation at a standard exercise intensity (Cochrane & Clark, 1990). Consequently, the patients reported lower ratings for breathlessness during exercise. In this study, these adaptations were not accompanied by improvement in indices of the severity of the underlying disease. Nevertheless, it can be argued that the benefit for the patient in terms of an increase in functional capacity and an improved response to sub-maximal exercise is worthwhile even in the absence of an effect on disease severity. If training does result in improvements in exercise induced asthma, as suggested by some (Svenonius, Kautto & Arborelius, 1983; Henrikkson & Nielsen, 1983), this is a bonus rather than the primary justification for the prescription of therapeutic exercise in this patient group.

Summary

The evidence that a moderate amount of physical activity can maintain or increase functional capacity in a large proportion of the population is convincing. Physical activity also contributes to the maintenance of desirable body weight, decreases the likelihood of some diseases and can help in the management of some patients with existing disease, contributing to an increased quality of life. The problem which urgently needs to be addressed is the prescription of physical activity appropriate to the circumstances and objectives of the individual. Current methods of doing this are relatively crude and justify further examination.

References

Baran, D.T., Kelly, A.M., Karellas, A., Gionet, M., Proce, M., Leahey, D., Steuterman, S., McSherry, B. & Roche, J. (1988). Ultrasonic attenuation of the os calcis in women with osteoporosis and hip fractures. *Calcified Tissue International*, **43**, 138–42.

Cade, R., Mars, D., Wagemaker, H., Zauner, C., Packer, D., Privette, M., Cade, M., Peterson, J. & Hood-Lewis, D. (1984). Effect of aerobic training on patients with systemic arterial hypertension. *American Journal of Medicine*, **77**, 785–90.

Cauley, J.A., Sandler, R., LaPort, R.E., Kriska, A., Hom, D. & Sashin, D. (1986). Physical activity and postmenopausal bone loss: results of a three year randomised clinical trial. *American Journal of Epidemiology*, **124**, 525.

Cavanagh, D.J. & Cann, C.E. (1988). Brisk walking does not stop bone loss in postmenopausal women. *Bone*, **9**, 201–4.

Chow, R., Harrison, J.E. & Notarius, C. (1987). Effect of two randomised exercise programmes on bone mass in healthy postmenopausal women. *British Medical Journal*, **295**, 1441–4.

Coats, J.S., Adamopoulos, S., Meyer, T.E., Conway, J. & Sleight, P. (1990). Effects of physical training in chronic heart failure. *Lancet*, **335**, 63–6.

Cochrane, L.M. & Clark A.J. (1990). Benefits and problems of a physical training programme for asthmatic patients. *Thorax*, **45**, 345–51.

Cook, T.C., Laporte, R.E., Washburn, R.A., Traven, N.D., Slemenda, C.W. & Metz, K.F. (1986). Chronic low level physical activity as a determinant of high density lipoprotein cholesterol and subfractions. *Medicine and Science in Sports and Exercise*, **18**, 653–7.

Dalen, N. & Olsson, N.E. (1974). Bone mineral content and physical activity. *Acta Orthopedica Scandinavica*, **45**, 170–4.

Danneskiold-Samsøe, B., Kofod, V., Munter, J., Grimby, G., Schnohr, P. & Jensen, G. (1984). Muscle strength and functional capacity in 78–81-year-old men and women. *European Journal of Applied Physiology*, **52**, 310–14.

Duncan, J.J., Farr, J.E., Upton, S.J., Hagan, R.D., Oglesby, M.E. & Blair, S.N. (1985). The effects of aerobic exercise on plasma catecholamines and blood pressure in patients with mild essential hypertension. *Journal of American Medical Association*, **254**, 2609–13.

Farrell, P.A. & Barboriak, J. (1980). The time course of alterations in plasma lipid and lipoprotein concentrations during eight weeks of endurance training. *Atherosclerosis*, **37**, 231–40.

Fleg, J.L. & Lakatta, E.G. (1988). Role of muscle loss in the age-associated reduction in $\dot{V}O_2max$. *Journal of Applied Physiology*, **65**, 1147–51.

Gollnick, P.D., Armstrong, R.B., Saltin, B., Saubert, C.W., Sembrowich, W.L. & Shepherd, R.E. (1973). Effect of training on enzyme activity and fiber composition of human skeletal muscle. *Journal of Applied Physiology*, **34**, 107–11.

Hagberg, J.M. (1987). Effect of training on the decline of $\dot{V}O_2max$ with aging. *Federation Proceedings*, **46**, 1830–3.

Hagberg, J.M., Montain, S.J., Martin, W.H. & Ehsani, A.A. (1989). Effect of exercise training in 60- to 69-year-old persons with essential hypertension. *American Journal of Cardiology*, **64**, 348–53.

Hardman, A.E., Hudson, A., Jones, P.R.M. & Norgan, N.G. (1988). Brisk walking influences the physiological responses to a submaximal step test in women. *Journal of Physiology*, **409**, 22P.

Hardman, A.E., Hudson, A., Jones, P.R.M. & Norgan, N.G. (1989). Brisk walking and plasma high density lipoprotein cholesterol concentration in previously sedentary women. *British Medical Journal*, **299**, 1204–5.

Hartung, G.H., Foreyt, J.P., Mitchell, R.E., Vlasek, I. & Gotto, A. (1980). Relation of diet to high-density lipoprotein cholesterol in middle-aged marathon runners, joggers, and inactive men. *New England Journal of Medicine*, **302**, 357–61.

Henrikkson, J.M. & Nielsen, T.T. (1983). Effects of physical training on exercise-induced broncho-constriction. *Acta Pediatrica Scandinavica*, **72**, 131–6.

Henrikkson, J & Reitman, J.S. (1977). Time course of changes in human skeletal muscle succinate dehydrogenase and cytochrome oxidase activities and maximal oxygen uptake with physical activity and inactivity. *Acta Physiologica Scandinavica*, **99**, 91–7.

Hickson, R.C., Bomze, H.A. & Holloszy, J.O. (1977). Linear increase in aerobic power induced by a strenuous program of endurance exercise. *Journal of Applied Physiology*, **42**, 372–6.

Hudson, A. & Hardman, A.E. (1989). Brisk walking as a means of improving aerobic fitness in middle-aged and older women. *Journal of Sports Sciences*, **7**, 66.

Huttunen, J.K., Lansimies, E., Voutilainen, E., Ehnholm, G., Hietanen, E., Penttila, I., Siitonen, O. & Rauramaa, R. (1979). Effect of moderate physical exercise on serum lipoproteins. A controlled clinical trial with special reference to serum high-density lipoproteins. *Circulation*, **60**, 1220–9.

Ingjer, F. (1979). Effects of endurance training on muscle fibre ATP-ase activity, capillary supply and mitochondrial content in man. *Journal of Physiology*, **294**, 419–32.

Jetté, M., Sidney, K. & Campbell, J. (1988). Effects of a twelve-week walking programme on maximal and submaximal work output indices in sedentary middle-aged men and women. *Journal of Sports Medicine and Physical Fitness*, **28**, 59–66.

Jones, P.R.M., Hardman, A.E., Hudson, A. & Norgan, N.G. (1991). Influence of brisk walking on the broadband ultrasonic attenuation of the calcaneus in previously sedentary women aged 30 to 61 years. *Calcified Tissue International*, **49**, 112–15.

Kiens, B. & Lithell, H. (1989). Lipoprotein metabolism influenced by training-induced changes in human skeletal muscle. *Journal of Clinical Investigation*, **83**, 558–64.

Krølner, B. & Toft, B. (1983). Vertebral bone loss: an unheeded side effect of therapeutic bedrest. *Clinical Science*, **64**, 537–40.

Krotkiewski, M. (1987). Can body fat patterning be changed? *Acta Medica Scandinavica*, Suppl. **723**, 213–23.

Lane, N.E., Bloch, D.A., Jones, H.H., Marshall, W.H., Wood, P.D. & Fries, J.F. (1986). Long distance running, bone density and osteoarthritis. *Journal of American Medical Association*, **255**, 1147–51.

Lanyon, L.E. (1990). Bone loading – the functional determinant of bone architecture and a physiological contributor to the prevention of osteoporosis. In: *Osteoporosis 1990*, Smith, R. ed., pp. 63–78. London: Royal College of Physicians.

Lapidus, L., Bengtsson, C., Larsson, B., Pennert, K., Rybo, E. & Sjostrom, L. (1984). Distribution of adipose tissue and risk of cardiovascular disease and death: a 12 year follow up of participants in the population study of women in Gothenburg, Sweden. *British Medical Journal*, **289**, 1257–61.

Maehlum, S., Grandmontagne, M., Newsholme, E.A. & Sejersted, O.M. (1986). Magnitude and duration of excess postexercise oxygen consumption in healthy young subjects. *Metabolism*, **35**, 425–9.

Marti, B., Suter, E., Riesen, W.F., Tschopp, A., Wanner, H-U. & Gutzwiller, F. (1990). Effects of long-term, self-monitored exercise on the serum lipoprotein and apolipoprotein profile in middle-aged men. *Atherosclerosis*, **81**, 19–31.

Martin, J.E. & Dubbert, P.M. (1982). Exercise applications and promotion in behavioural medicine: current status and future directions. *Journal of Consulting and Clinical Psychology*, **50**, 1004–17.

Meredith, C.N., Frontera, W.R., Fisher, E.C., Hughes, V.A., Herland, J.C., Edwards, J. & Evans, W.J. (1989). Peripheral effects of endurance training in young and old subjects. *Journal of Applied Physiology*, **66**, 2844–9.

Miller, G.J. & Miller, N.E. (1975). Plasma-high-density-lipoprotein concentration and development of ischaemic heart disease. *Lancet*, **i**, 16–19.

Morris, J.N., Chave, S.P.W., Adam, C., Sirey, C., Epstein, L. & Sheehan, D.S. (1973). Vigorous exercise in leisure time and the incidence of coronary heart disease. *Lancet*, **i**, 333–9.

Morris, J.N., Clayton, D.G., Everitt, M.G., Semmence, A.M. & Burgess, E.H. (1990). Exercise in leisure time: coronary attack and death rates. *British Heart Journal*, **63**, 325–34.

Morris, J.N., Everitt, M.G., Pollard, R. & Chave, S.P.W. (1980). Vigorous exercise in leisure time: protection against coronary heart disease. *Lancet*, **ii**, 1207–10.

Morris, J.N., Heady, J.A., Raffle, P.A.B., Roberts, C.G. & Parks, J.W. (1953). Coronary heart disease and physical activity of work. *Lancet*, **ii**, 1053–7, 1111–20.

Nelson, L., Jennings, G.L., Esler, M.D. & Korner, P.I. (1986). Effect of changing level of physical activity on blood-pressure and haemodynamics in essential hypertension. *Lancet*, **ii**, 473–6.

O'Connor, G.T., Buring, J.E., Yusuf, S., Goldhaker, S.Z., Olmstead, E.M., Paffenbarger, R.S. & Mennekens, C.H. (1989). An overview of randomized trials of rehabilitation with exercise after myocardial infarction. *Circulation*, **80**, 234–44.

Paffenbarger, R.S. & Hale, W.E. (1975). Work activity and coronary heart mortality. *New England Journal of Medicine*, **292**, 545–50.

Paffenbarger, R.S., Hyde, R.T., Wing, A.L. & Hsieh, C. (1986). Physical activity, all-cause mortality and longevity of college alumni. *New England Journal of Medicine*, **314**, 605–13.

Paffenbarger, R.S., Wing, A.L. & Hyde, R.T. (1978). Physical activity as an index of heart attack risk in college alumni. *American Journal of Epidemiology*, **108**, 161–75.

Pavlou, K.N., Krey, S. & Steffee, W.P. (1989). Exercise as an adjunct to weight loss and maintenance in moderately obese subjects. *American Journal of Clinical Nutrition*, **49**, 1115–23.

Pavlou, K.N., Whatley, J.E., Jannace, P.W., DiBartolomeo, J.J., Burrows, B.A., Duthie, E.A.M. & Lerman, R.H. (1989). Physical activity as a supplement to a weight-loss dietary regimen. *American Journal of Clinical Nutrition*, **49**, 1110–14.

Pirnay, F., Bodeux, M., Criwelaard, J.M. & Franchimont, P. (1987). Bone mineral content and physical activity. *International Journal of Sports Medicine*, **8**, 331–5.

Pollock, M.L., Miller, H.S., Janeway, R., Linnerud, A.C., Robertson, B. & Valentino, R. (1971). Effects of walking on body composition and cardiovascular function in middle-aged men. *Journal of Applied Physiology*, **30**, 126–30.

Porcari, M.S., McCarron, R., Kline, G., Freedson, P.S., Ward, A., Ross, J.A. & Rippe, J.M. (1987). Is fast walking an adequate aerobic training stimulus for 30- to 69-year-old men and women? *Physician and Sportsmedicine*, **15**, 119–29.

Powell, K.E., Thompson, P.D., Casperson, C.J. & Kendrick, J.S. (1987). Physical activity and the incidence of coronary heart disease. *Annual Review of Public Health*, **8**, 253–87.

Rechnitzer, P.A., Cunningham, D.A., Andrew, G.M., Buck, C.W., Jones, N.L., Kavanagh, T., Oldridge, N.B., Parker, J.O., Shephard, R.J., Sutton, J.R. &

Donner, A.P. (1983). Relation of exercise to the recurrence rate of myocardial infarction in men. *American Journal of Cardiology*, **51**, 65–9.

Rubin, C.T. & Lanyon, L.E. (1984). Regulation of bone formation by applied dynamic loads. *Journal of Bone and Joint Surgery*, **66A**, 397–402.

Saltin, B., Blomqvist, G., Mitchell, J.H., Johnson, R.L., Wildenthal, K. & Chapman, C.B. (1968). Response to exercise after bed rest and training. *Circulation*, **38** (Suppl. 7), 1–78.

Santiago, M.C., Alexander, J.F., Stull, G.A., Serfass, R.C., Hayday, A.M. & Leon, A.S. (1987). Physiological responses of sedentary women to a 20-week conditioning program of walking or jogging. *Scandinavian Journal of Sports Science*, **9**, 33–9.

Seals, D.R., Hurley, B.F., Schultz, J. & Hagberg, J.M. (1984). Endurance training in older men and women. II. Blood lactate response to submaximal exercise. *Journal of Applied Physiology*, **57**, 1030–3.

Shephard, R.J. (1986). *Fitness of a Nation*. In Medicine and Sport Science. Hebbelink, M. and Shephard, R.J., eds, vol. 22, Basel: Karger.

Simkin, A., Ayalon, J. & Leichter, I. (1987). Increased trabecular bone density due to bone-loading exercises in postmenopausal osteoporotic women. *Calcified Tissue International*, **40**, 59–63.

Stensel, D.J., Hardman, A.E., Brooke-Wavell, K., Jones, P.R.M. & Norgan, N.G. (1992). The influence of brisk walking on endurance fitness in previously sedentary, middle-aged men. *Journal of Physiology*, 446, 123P.

Stock, M. & Rothwell, N. (1982). *Obesity and Leanness: Basic Aspects*. London: Libbey.

Stubbe, I., Hansson, P., Gustafson, A. & Nilsson-Ehle, P. (1983). Plasma lipoproteins and lipolytic enzyme activities during endurance training in sedentary men: changes in high-density lipoprotein subfractions and composition. *Metabolism*, **32**, 1120–8.

Stupakov, G.P., Kazaykin, V.S., Kozlovsky, A.O. & Korolev, V.V. (1984). Evaluation of changes in human axial skeletal bone structures during long-term space flights. *Kosm Biol Aviakosm Medicine*, **18**, 33.

Svenonius, E., Kautto, R. & Arborelius, M. (1983). Improvement after training of children with exercise-induced asthma. *Acta Pediatrica Scandinavica*, **72**, 23–30.

Thompson, P.D., Lazarus, B., Cullinane, E., Henderson, L.O., Musliner, T., Eshleman, R. & Herbert, P.N. (1983). Exercise, diet, or physical characteristics as determinants of HDL levels in endurance athletes. *Atherosclerosis*, **46**, 333–9.

Tremblay, A., Despres, J-P., Leblanc, C., Craig, C.L., Ferris, B., Stephens, T. & Bouchard, C. (1990). Effect of intensity of physical activity on body fatness and fat distribution. *American Journal of Clinical Nutrition*, **51**, 153–7.

Welsh Heart Programme Directorate (1987). *Exercise for Health: Health Related Fitness in Wales*. Heartbeat Report No. 23.

Wenger, N.K. (1989). Rehabilitation of the patient with coronary heart disease. *Postgraduate Medicine*, **85**, 369–80.

White, M.K., Martin, R.B., Yeater, R.A., Butcher, R.L. & Radin, E.L. (1984). The effects of exercise on the bones of postmenopausal women. *International Orthopaedics*, **7**, 209–14.

Williams, P.T., Krauss, R.M., Wood, P.D., Lindgren, F.T., Giotas, C. & Vranizan, K.M. (1986). Lipoprotein subfractions of runners and sedentary men. *Metabolism*, **35**, 45–52.

Woo, R., Garrow, J.S. & Pi-Sunyer, F.X. (1982). Voluntary food intake during prolonged exercise in obese women. *American Journal of Clinical Nutrition*, **36**, 478–84.

Wood, P.D., Haskell, W.L., Blair, S.N., Williams, P.T., Krauss, R.M., Lindgren, F.T., Albers, J.J., Ho, P.H. & Farquhar, J.W. (1983). Increased exercise level and plasma lipoprotein concentrations: a one-year, randomized, controlled study in sedentary, middle-aged men. *Metabolism*, **32**, 31–9.

Wood, P.D., Stefanick, M.L., Dreon, D.M., Frey-Hewitt, B., Garay, S.C., Williams, P.T., Superko, H.R., Fortmann, S.P., Albers, J.J., Vranizan, K.M., Ellsworth, N.M., Terry, R.B. & Haskell, W.L. (1988). Changes in plasma lipids and lipoproteins in overweight men during weight loss through dieting as compared with exercise. *New England Journal of Medicine*, **319**, 1173–9.

Wood, P.D., Terry, R.B. & Haskell, W.L. (1985). Metabolism of substrates: diet, lipoprotein metabolism and exercise. *Federation Proceedings*, **44**, 358–63.

Zylstra, S., Hopkins, A., Erk, M., Hreshchyshyn, M.M. & Anbar, M. (1989). Effect of physical activity on lumbar spine and femoral neck bone densities. *International Journal of Sports Medicine*, **10**, 181–6.

12 *Physical activity, obesity and weight maintenance*

WIM H.M. SARIS

Introduction

Obesity is recognized as a major health problem in western society. Maintenance of an acceptable body weight is therefore an important aspect of a healthy lifestyle.

Overweight and obesity result from a state of positive energy balance in which total energy intake from food has exceeded expenditure over a prolonged period. Although this equation of energy balance is simple, it has not been proved which one of the two factors, energy in or energy out, is responsible for the imbalance and the consequent high energy stores. There are many factors associated with the energy intake and expenditure and the interaction of these two factors which makes this equation complex. Furthermore, the positive balance is a gradual process over many years. In the longitudinal Zutphen study of middle-aged men over 25 years, a significant increase of 2.8 kg was noticed (Kromhout *et al.*, 1990). Assuming this to be an increase in fat stores, suggests an imbalance of about 70 MJ or 8 kJ per day. With the most precise methods that are available today, this imbalance is not measurable. Even if the weight increase in the upper three percentiles of this cohort of about 15 kg is considered, the imbalance on a daily basis is about 50 kJ. The precision of a 24 h measurement of energy expenditure in a respiration chamber is at best in the order of 2 to 3% leading to absolute differences of 100–150 kJ. In such a situation, physical activity is restricted to a small room, and may therefore hide the possible cause of imbalance. The measurement under free living conditions with the new doubly labelled water technique over a period of 2 weeks has a precision of 4–7% (Schoeller, 1988). Even with this technique, the described theoretical differences in energy expenditure are difficult to identify.

With respect to energy intake, the situation is even worse. Comparison of self-reported energy intake with energy expenditure measured by the doubly labelled water technique has shown that errors in energy intake are large (Schoeller, 1990). Reported intakes tend to be lower than expenditure and available data show a gross underreporting with higher body mass index.

These findings have major implications for the interpretation of individual energy balance, leading to the conclusion that even the best measures of energy input and output are too variable to be useful to study energy balance under free living conditions.

Components of energy expenditure

Energy expenditure over 24 h can be divided into four components as described by Horton (1983). These components are basal or resting metabolic rate (BMR or RMR); thermic effect of food (TEF); thermic effect of activity (TEA) and adaptive thermogenesis (AT). RMR may account for 65–70% of the daily energy expenditure. A number of factors such as age, sex, body composition and genetic background influence RMR.

The genetic contribution to the individual variation of RMR has attracted particular attention since recent publications showed strong evidence for a genetic component in the origin of obesity (Bogardus *et al.*, 1986; Stunkard *et al.*, 1990; Bouchard *et al.*, 1990). Such differences in genetic expression could have a significant influence on the ability to maintain energy balance throughout life. TEF comprises two components, termed obligatory and facultative thermogenesis. The latter is of special interest in relation to physical activity and maintenance of body weight. It may have its origin in the increased sympathetic nervous system activity (Sims & Danforth, 1987).

The most effective way to increase total energy expenditure is to increase TEA which may account for as little as 15%, or as much as 60% of total energy expenditure. As will be discussed later on, changes in TEA may also influence RMR and TEF.

The least known component in human energy expenditure is AT. It is illustrated by the adaptive decrease in RMR during starvation in the classical studies of Keys *et al.* (1950). Recent studies of Tremblay *et al.* (1988) showed the same phenomenon of decreased TEF in highly trained athletes. Facultative thermogenesis might also be considered as a form of AT.

Although the changes in TEE as a result of AT are at most 10–15%, such changes might have a decisive impact on energy balance in the long term.

The effects of exercise on energy expenditure

It is attractive to postulate that inactivity may be the key factor in the development of a positive energy balance leading to obesity. A plethora of studies have examined the question of inactivity in obese subjects. Although available data are by no means conclusive, it appears that in the majority of studies cited in a review concerning this topic, no relationship was found (Pacy, Webster & Garrow, 1986). Six studies reported lower activity levels in obese subjects, but ten failed to find a significant difference.

One of the major problems with respect to these studies is the measurement and definition of daily physical activity. Reduced levels of activity may be balanced partly by the increased energy cost of weight-bearing activities. A lower level of movement, for instance, a smaller walking distance during a ball game, does not automatically mean less TEE. In fact, moving around with a higher body mass implies a higher energy cost. This was elegantly demonstrated in a study by Weigle (1988), where, during a weight-reducing regime, the experimental subjects wore a vest with weights compensating for body weight loss. The decrease of TEE was only 50% compared to the control obese group with identical weight losses. Also, in the study of Waxman & Stunkard (1980) differences in TEE between obese and lean children disappear when corrections were made for the larger body mass, despite the fact that observation scores of daily physical activity were significantly lower in the obese children.

Recently, the first results were published comparing obese and non-obese adolescents using the doubly labelled water method, nowadays considered as the golden standard method for measuring TEE under free living conditions over a period of 14 days (Bandini, Schoeller & Dietz, 1990). No differences were found in TEE expressed as multiples of RMR (1.79 and 1.68 in non-obese/obese males respectively and 1.69 and 1.74 in non-obese/obese females respectively).

In general, one can conclude that, although the data are conflicting, there is growing evidence that TEA of the obese is not decreased, although on observation obese subjects do give the impression of being less active. Perhaps only in the situation of extreme obesity is physical activity so affected that it lowers TEA.

In addition to the direct effect of physical activity on TEA, exercise may influence TEE by means of a prolonged stimulating effect on RMR. This possible increased metabolic response of exercise, also termed post-exercise oxygen consumption (EPOC), is now under debate. Recently Poehlman, Christopher & Gorman (1991) reviewed the literature on this matter. Studies have shown that the duration of EPOC may vary from less than 1 to 24 h post-exercise, increasing RMR from between 1% and 25%. They concluded that the greater the exercise perturbation the greater the magnitude of the EPOC. However, considering that the exercise prescription for the general population consists of low to moderate exercise, the effects of EPOC do not produce a substantial contribution to the TEE

(15–300 kJ). On the other hand, such an increase in energy expenditure together with the costs of the exercise itself, could be of value in the long term in maintaining energy balance.

Besides this EPOC effect, it is suggested that regular exercise may lead to a higher RMR as a consequence of a better training status leading to a higher aerobic power. Poehlman *et al.* (1989) hypothesized that RMR could be increased when the energy turnover rate is high, concurrent with the maintenance of energy balance. In support of this hypothesis is the observation of Tremblay *et al.* (1988) showing a 6.6% drop in RMR of highly trained subjects following 3 days of detraining.

If there is an effect of training status on RMR, one should also expect an effect of the level of physical activity on RMR due to the fact that a higher level of daily physical activity will increase aerobic fitness. In general, cross-sectional studies have failed to provide support for this view. Poehlman *et al.* (1991) suggested that differences in experimental design and especially the lack of precise methods to measure daily physical activity, may have obscured such a relationship. In a recent study by Westerterp *et al.* (1991) using the doubly labelled water method to measure daily physical activity in sedentary and moderately active subjects, a positive relationship between TEA and RMR was demonstrated ($r^2 = 0.72$). This observation from the use of the best possible technique, underlines the remarks of Poehlman and co-workers. The subjects in this study were definitely not involved in any kind of training. Nevertheless, a strong relation between daily physical activity and RMR was demonstrated.

Another factor that might play a role of importance in the level of TEE in relation to exercise is the compensation in daily activities outside the training hours leading to no extra increase in 24 h. This question was studied in two adult groups and in a group of 10 year-old boys (Van Dale *et al.*, 1989; Meyer *et al.*, 1991; Blaak *et al.*, 1992). In lean and obese women, no indication of compensation in normal daily activity was observed. In lean males, training stimulated physical activity during the non-exercising part of the day. In the 10 year-old obese boys, a normal daily physical activity level was observed compared to their lean counterparts. As a result of a training programme, TEE was increased by 10%. About 50% of the extra expenditure could be accounted for by the training programme itself. It was concluded that other energy-consuming factors such as TEF, EPOC or TEA outside the training hours must have been increased. Given the absolute value of this 50% increment, it is clear that TEA outside the training hours is the first candidate for this remarkable extra increase in energy expenditure. One can conclude from these studies that there are no indications of a compensation behaviour during these training programmes.

In the study of human energy expenditure, one of the more controversial issues is whether obese subjects have a defective thermogenic response to food in comparison to their lean counterparts, and how far exercise can potentiate the effect on TEF. A wide variety of answers to these questions has been found. A salient finding of these studies is the relatively low reproducibility of the measurement of TEF. Intra-individual coefficients of variation are around 30% (Ravussin *et al.*, 1986). This may pose serious problems for studies that aim to determine within-person responses to the potentiating effect of exercise. In addition, lack of standardized protocols also contribute significantly to the existing confusion.

A final interesting observation about the effectiveness of exercise on energy balance is the sex difference. In a study on the effects of a 5-month endurance training programme, Meyer *et al.* (1991) showed a marked gender difference in the metabolic response. The metabolic response to exercise was much larger in men than in women, and exceeded the net costs of the exercise itself. They suggested that in men, exercise leads to a stimulation of TEA and TEF, whereas women do not show such a response. Also different changes in body composition and adipose tissue function was noticed in a 20-week aerobic exercise programme (Tremblay *et al.*, 1984). Males demonstrated a significant reduction in body weight, fat mass and fat cell weight, while females demonstrated no significant changes in these parameters. Epinephrine-stimulated lipolysis was considerably more pronounced in males (+66%) than in females (+46%) (Despres *et al.*, 1984). Differences in distribution of adipose tissue may be the key factor to this different response. Male subjects are characterized by a storage capacity in the abdominal region. The abdominal fat cell has a more sensitive release and storage of triglycerides. Leibel *et al.* (1985) found a higher number of beta receptors in abdominal fat cells that were more responsive to exercise. In women, storage capacity is located in the femoral region, particularly where alpha receptors are predominant. Higher levels of both basal and stimulated lipolysis in abdominal fat cells have been observed, denoting an increased fat oxidation capacity during exercise. This leads to the recent hypothesis that overweight is related to the capacity to oxidize fat (Zurlo *et al.*, 1990*a*). It also links the obesity problem to the metabolic potentials of the muscle cell. Recently, Wade, Marbet & Round (1990) demonstrated that obese men combusted less fat during moderate work than lean men. They also found that the proportion of slow muscle fibres was inversely related to fatness. Slow (type 1) muscle fibres which are well endowed with mitochondria, usually work oxidatively and use fatty acids as an important fuel source, especially during low and moderate exercise. In the study of Zurlo *et al.* (1990*b*) 24 h respiratory exchange ratio (RER) was related significantly to the subsequent changes in body weight.

Subjects with a higher 24 h RER had a 2.5 times higher risk of gaining more than 5 kg body weight compared to subjects with a low 24 h RER. Astrup, Bülow & Madsen (1985) demonstrated that about 50% of the increase in oxygen consumption induced by ephedrine may take place in skeletal muscle. In addition, Zurlo *et al.* (1990*a*) showed that the differences in energy expenditure in skeletal muscle partly explained the variance in RMR and TEE among subjects.

Therefore, several factors might influence the variation in muscle energy metabolism along with fibre type and sympathetic–adrenergic response. Both are effectively influenced by training. Longitudinal and cross-sectional studies with animals and man have demonstrated dramatic changes in the skeletal muscle in response to regular exercise (Saltin & Gollnick, 1983). The concentration of mitochondrial protein and the capacity for several metabolic pathways including beta-oxidation is enhanced, leading to a greater use of lipids during sub-maximal exercise. Furthermore, in relation to the sympathetic–adrenergic control, endurance training augments the stimulatory effect of ephedrine on energy metabolism in the skeletal muscle (Richter *et al.*, 1984). The proposed mechanism responsible for this increased energy turnover is substrate cycling (interconversion of metabolic intermediates which allows for greater enzyme sensitivity) as proposed by Newsholme (1980). He suggested that before, during and after exercise, the rate of cycling is increased. It is likely that the capacity and responsiveness of these cycles to regulatory hormones are increased by training.

Dietary quality and exercise

Man has only a limited capacity to convert excess dietary carbohydrate to fat via de novo fatty acid synthesis. As a result, high intakes of carbohydrate induce maximization of the glycogen stores and increased resting glucose oxidation. This increase in carbohydrate oxidation is mediated by an increase in sympathetic nervous system activity. Although the storage capacity for carbohydrate is limited (about 600 g), the metabolic cost for storage of carbohydrate is considerably less than converting it to fat. On the other hand, the body stores incorporate ingested fat much more efficiently (approximately 3% of the ingested energy) than when carbohydrate is converted into triglyceride (approximately 23%). As a consequence, excess intake of fat will increase body fat stores more efficiently than carbohydrate. Abbott *et al.* (1988) reported a significant correlation between fat balance and energy balance, suggesting that excess fat intake is stored as excess energy, while excess carbohydrate and protein are closely regulated by changes in oxidation rate.

It was Flatt (1978) who postulated that the food fat/carbohydrate quotient (FQ) must parallel the fat/carbohydrate fuel oxidation quotient (RQ). Excess fat intake (low FQ) may result in increased food intake in order to achieve carbohydrate balance leading to an overall energy intake surplus. Although the results from epidemiological data are not convincing, owing to the methodological problems of getting valid information about food intake, it has been shown that high fat intake is related to an increase of body weight (Lissner *et al.*, 1987) and that high fat consumers have increased body fat mass (Tremblay *et al.*, 1989).

While still the topic of some debate, it appears that fat stores are not subject to accurate metabolic control in contrast to carbohydrate and protein. It is suggested therefore that, in particular, the improvement of the dietary carbohydrate/fat ratio is one of the factors that can prevent weight gain. A possible route by which exercise can lead to an improvement of the dietary intake is the observed change to higher carbohydrate intake at the expense of fat in physically active subjects (van Erp-Baart *et al.*, 1989). In a group of 419 athletes, participating in different types of sports at international level, the relative intake of carbohydrate was positive related to the energy intake. This is not surprising in the light of well-established knowledge of an enhanced need of carbohydrate to perform at a maximal level. However, even in the athletic groups with a relatively low level of energy intake, the percentage of carbohydrate was higher than in the general population. Although maximal performance is not the ultimate goal of exercise therapy in obesity, the improvement of the carbohydrate/fat ratio in combination with a higher activity level (moderate endurance exercise) to increase fat oxidation may be of benefit in the prevention of weight gain.

Effect of exercise on success of weight maintenance

Many obese subjects claim that a life-long diet or extreme exercise levels are needed to maintain weight loss. In general, one can say that the long-term results on weight maintenance are very disappointing. These poor results have led to speculations about a post-obesity syndrome of a persistently lower level of energy expenditure. Studies have shown that post-obese subjects have enhanced metabolic efficiency. In other studies, however, a similar metabolic rate for successful reduced obese and lean subjects has been observed (Saris, 1989). A group of obese women were followed over a period of 18 to 40 months post-treatment (van Dale, Saris & Hoor ten., 1990). Like other investigations, it was observed that the maintenance of body weight after treatment turned out to be very difficult. In contrast to these disappointing results, a better weight maintenance was observed in a small number of subjects (13%) who had regular exercise (three times or more per week) after the weight loss period. These subjects maintained most of their weight loss. Remarkably, RMR per kg lean body mass (LBM) was still about 10% lower in the unsuccessful group 18 to 40 months after the energy restriction period compared to the starting values before treatment. This is in contrast to the physically active group who restored RMR per kg LBM to the initial levels. This persistent depression of RMR will have its effect on TEE, since RMR accounts for about 75% of TEE.

Finally, from the reviewed research, it seems that exercise plays an important role in the prevention of weight gain. Increasing physical activity seems to be an effective way to increase the metabolic potential to maintain energy balance. Several factors are involved that are not fully understood. Nevertheless, all available data indicates that exercise is one of the most powerful effectors in this complex and delicate balance between energy input and output.

References

Abbott, W.G.H., Howard, B.V., Christin, L., Freymond, S., Lillioja, S., Boyce, V.L., Anderson, T.E., Bogardus, C. & Ravussin, E. (1988). Short term energy balance: relationship with protein, carbohydrate and fat balance. *American Journal of Physiology*, **225**, E332–7.

Astrup, A., Bülow, J. & Madsen, J. (1985). Contribution of BAT and skeletal muscle to thermogenesis induced by ephedrine in man. *American Journal of Physiology*, **248**, E507–14.

Bandini, L.G., Schoeller, D.A. & Dietz, W.H. (1990). Energy expenditure in obese and non-obese adolescents. *Pediatric Research*, **27**, 198–203.

•Blaak, E.E., Westerterp, K.R., Bar-or, O., Wouters, L.J.M. & Saris, W.H.M. (1992). Effect of training on total energy expenditure and spontaneous activity in obese boys. *American Journal of Clinical Nutrition*, **55**, 777–89.

Bogardus, C., Lillioja, S., Ravussin, E., Abbot, W., Zawadzki, J.K., Young, A., Knowler, W.C., Jacobowitz, R. & Moll, P.P. (1986). Familial dependence of the resting metabolic rate. *New England Journal of Medicine*, **315**, 96–100.

Bouchard, C., Tremblay, A., Depres, J.P., Nadeau, A., Lupien, P.J., Theriault, G., Dussault, J., Moorjani, S., Pinault, S. & Fournier, G. (1990). The response to long term overfeeding in identical twins. *New England Journal of Medicine*, **322**, 1477–82.

Dale van, D., Schoffelen, P.F.M., Hoor ten, F. & Saris, W.H.M. (1989). Effect of adding exercise to energy restriction, 24-h energy expenditure, resting metabolic rate and daily physical activity. *European Journal of Clinical Nutrition*, **43**, 441–51.

Dale van, D., Saris, W.H.M. & Hoor ten, F. (1990). Weight maintenance and resting metabolic rate 18–40 months after a diet/exercise treatment. *International Journal of Obesity*, **14**, 347–59.

Despres, J.P., Bouchard, C., Savard, R., Tremblay, A., Marcotte, M. & Theriault, G. (1984). The effect of a 20 weeks endurance training programme on adipose tissue morphology and lipolysis in men and women. *Metabolism*, **33**, 235–9.

van Erp-Baart, A.M.J., Saris, W.H.M., Binkhorst, R.A., Vos, J.A. & Elvers, J.W.H. (1989). Nationwide study on nutritional habits in elite athletes. Part I. Energy carbohydrates, protein and fat intake. *International Journal of Sports Medicine*, **10**, S3–10.

Flatt, J.P. (1978). The biochemistry of energy expenditure. In *Recent Advances in Obesity Research II*. Bray, G.A., ed., pp. 211–28, London: Newman Publications.

Horton, E.S. (1983). Introduction: an overview of the assessment and regulation of energy balance in humans. *American Journal of Clinical Nutrition*, **38**, 972–7.

Keys, A., Brozek, J., Henshel, A., Michelson, O. & Taylor, H.L. (1950). *The Biology of Human Starvation*, vol. 1. Minneapolis: University of Minneapolis Press.

Kromhout, D., Lezenne Coulander de, C., Obermann-Boer de, G.L., Kampen-Donker van, M., Goddijn, E. & Bloemberg, B.P.M. (1990). Changes in food and nutrient intake in middle-aged men from 1960 to 1985. The Zutphen Study. *American Journal of Clinical Nutrition*, **51**, 123–9.

Leibel, R.L., Hirsch, J., Berry, E. & Green, R.K. (1985). Alterations in adipocyte free fatty acid reesterification associated with obesity and weight reduction in man. *American Journal of Clinical Nutrition*, **42**, 198–206.

Lissner, L., Levitsky, D.A., Strupp, B.J., Kalwarf, H.J. & Roe, D.A. (1987). Dietary fat and the regulation of energy intake in human subjects. *American Journal of Clinical Nutrition*, **46**, 886–92.

Meyer, G.A.L., Janssen, G.M.E., Westerterp, K.R., Verhoeven, F., Saris, W.H.M. & Hoor ten, F. (1991). The effect of a 5 months training program on physical activity: evidence for a sex difference in the metabolic response to exercise. *European Journal of Applied Physiology*, **62**, 11–17.

Newsholme, E.A. (1980). A possible metabolic basis for the control of body weight. *New England Journal of Medicine*, **302**, 400–5.

Pacy, P.J., Webster, J. & Garrow, J.S. (1986). Exercise and obesity. *Sports Medicine*, **3**, 89–113.

Poehlman, E.T., Melby, C.L., Badylak, S.F. & Calles, J. (1989). Aerobic fitness and resting energy expenditure in young adult males. *Metabolism*, **38**, 85–90.

Poehlman, E.T., Christopher, L.M. & Gorman, M.I. (1991). The impact of exercise and diet restriction daily energy expenditure. *Sports Medicine*, **2**, 78–101.

Ravussin, E., Lillioja, S., Anderson, T.E., Christin, L. & Bogardus, C. (1986). Determinants of 24-hours energy expenditure in man. Methods and results using a respiration chamber. *Journal of Clinical Investigation*, **78**, 1568–78.

Richter, E.A., Christensen, N.J., Ploug, T. & Galbo, H. (1984). Endurance training augments the stimulatory effect of epinephrine on oxygen consumption in perfused skeletal muscle. *Acta Physiologica Scandinavica*, **120**, 613–15.

Saltin, B. & Gollnick, P.D. (1983). Skeletal muscle adaptability. Significance for metabolism and performance. In *Handbook of Physiology: Skeletal Muscle*. Peach, L.D., Adrian, R.H. & Geiger, S.R., eds, Baltimore: Williams and Wilkins.

Saris, W.H.M. (1989). Physiological aspects of exercise in weight cycling. *American Journal of Clinical Nutrition*, **49**, 1099–104.

Schoeller, D.A. (1988). Measurement of energy expenditure in free living humans by using doubly labeled water. *Journal of Nutrition*, **118**, 1278–89.

Schoeller, D.A. (1990). How accurate is self-reported dietary energy intake? *Nutrition Reviews*, **48**, 373–9.

Sims, E.A.H. & Danforth, E. (1987). Expenditure and storage of energy in man. *American Journal of Clinical Nutrition*, **79**, 1019–25.

Stunkard, A.J., Harris, J.R., Pedersen, N.L. & McClean, G.E. (1990). The body mass index of twins who have been reared apart. *New England Journal of Medicine*, **322**, 1483–7.

Tremblay, A., Despres, J.P., Leblanc, C. & Bouchard, C. (1984). Sex dimorphism in fat loss in response to exercise training. *Journal of Obesity and Weight Regulation*, **3**, 193–203.

Tremblay, A., Nadenau, A., Fournier, G. & Bouchard, C. (1988). Effect of a three days interruption of exercise training on RMR and GIT in trained individuals. *International Journal of Obesity*, **12**, 163–8.

Tremblay, A., Piourde, G., Despres, J.P. & Bouchard, C. (1989). Impact of dietary fat content and fat oxidation on energy intake in humans. *American Journal of Clinical Nutrition*, **49**, 799–805.

Wade, A.J., Marbet, M.M. & Round, J.M. (1990). Muscle fiber type and etiology of obesity. *Lancet*, **355**, 805–8.

Waxman, M. & Stunkard, A.J. (1980). Caloric intake and expenditure of obese boys. *Journal of Pediatrics*, **96**, 187–93.

Weigle, D.S. (1988). Contribution of decreased body mass to diminish thermic effect of exercise in reduced-obese men. *International Journal of Obesity*, **12**, 567–78.

Westerterp, K.R., Saris, W.H.M., Soeters, P.B., Winants, Y. & Hoor ten, F., (1991). Physical activity and sleeping metabolic rate. *Medicine and Science in Sport and Exercise*, **23**, 166–70.

Zurlo, F., Larson, K., Bogardus, C. & Ravussin, E. (1990*a*). Skeletal muscle metabolism is a major determinant of resting energy expenditure. *Journal of Clinical Investigation*, **86**, 1423–7.

Zurlo, F., Lillioja, S., Puente, A., Nyomba, B.L., Raz, I., Saad, M.F., Swinburn, B.A., Knowler, W.C., Bogardus, C. & Ravussin, E. (1990*b*). Low ratio of fat to carbohydrate oxidation as predictor of weight gain: Study of 24 hour RQ. *American Journal of Physiology*, **259**, E650–7.

13 *Adherence to physical activity and exercise*

STUART BIDDLE

Documentation of the potential physical and psychological effects of participation in physical activity and exercise is now quite extensive (Bouchard *et al.*, 1990). However, despite a widespread awareness of these benefits, population estimates for industrialized nations suggest that involvement in physical activity and exercise that is likely to produce cardiovascular benefit remains a minority pursuit (Stephens, Jacobs & White, 1985). The prevalence of involvement in activity levels that might impact on other aspects of fitness or wider parameters of physical health remains unknown, as does the level of involvement that affects mental health. It is considered a research priority, therefore, to identify the factors influencing the adoption and maintenance of physical activity and exercise. This is usually termed 'exercise adherence' (Biddle & Fox, 1989; Dishman, 1988*a*).

The purpose of this chapter is to summarize the current knowledge on the likely psychological correlates or proposed 'determinants' of participation in physical activity and exercise. Physical activity (PA) will be used to refer to all muscular–skeletal movement that results in energy expenditure, thus being an all-encompassing category including exercise and sport, as well as low-intensity movement not usually associated with overt gains in 'fitness'. Exercise, however, will be used to refer to structured forms of PA that are usually participated in for reasons of gaining, maintaining or improving fitness (Caspersen, Powell & Christenson, 1985), and therefore is often of a moderately vigorous or vigorous nature. A distinction between the two terms will be used where appropriate or where evidence or experience dictates.

The chapter will consider independently the factors thought likely to influence children and adults. A cognitive–developmental perspective is adopted for the discussion on children, and the heuristic model proposed by Sallis & Hovell (1990) will be adopted for discussing the determinants of involvement of adults in PA and exercise.

Children

The activity levels of adults, as stated, would appear to be low from the standpoint of cardiovascular benefit (Stephens, Jacobs & White, 1985). Similarly, limited evidence on children suggests that they do not engage in sustained exercise likely to benefit the cardiovascular system, even though they are probably the 'fittest' section of society (Armstrong *et al.*, 1990*a*, *b*). However, while physical activity levels in adults are known to be associated with coronary heart disease (CHD) risk, there is no evidence for this in children, other than the possibility that inactive children are storing up problems for the future (Montoye, 1986). This suggests that the issue to be addressed is not necessarily how to get children fit and active in the short term, but how to help children become active adults. It is possible that some strategies, designed for the former, may be counter-productive for the latter.

The few studies that have looked at the tracking of sports and exercise from childhood to adulthood have shown mixed results (Engstrom, 1986; Powell & Dysinger, 1987). The notion of an 'activity habit' is not well supported, although Engstrom (1986) suggests that the perceptions children hold about PA might be a more fruitful line of enquiry, particularly when put alongside influences of the environment (Fox, 1991). Similarly, younger children will not have the cognitive maturity to have 'health', or even 'fitness', as goals for participation. Developmental psychologists have suggested that children's conceptions of health and illness follow closely the Piagetian stages of preoperational, concrete operational, and formal operational thought (Maddux *et al.*, 1986; Natapoff, 1982). The move through such stages contributes to the child being able to operate in a more autonomous way with perceptions of greater control over their own health and illnesses. Whether such a model is applicable to exercise as a health behaviour has not been tested. Nevertheless, it has been suggested that children at the stage of concrete operational thought are able to assume some behavioural responsibility for their own health by avoiding sources of illness. This might equate to children taking part in PA and exercise as 'preventive health'. However, whether such long-term goal-setting is evident in children is doubtful. Similarly, more advanced children, at the formal operational stage of thinking, may be able to perceive the multiple causes and cures of health and illness. Bibace & Walsh (1979) label this the 'psychophysiological stage' of children's health conceptualizations and suggest that, at this stage, children are able to understand, albeit at a simple level, that there is a relationship between thoughts, feelings and physical processes of the body, and between stress and ill health.

Maddux *et al.* (1986), however, point out that using adult models of health decision-making may be inappropriate for the study of health behaviours in children. Such theories, they say, are 'based on the assumption that people are capable of gathering information about health and behaviour and of arriving at rational and logical decisions about health-related behaviours based on expectancies, probabilities, reinforcement values, cost–benefit ratios, and other similar factors. Despite the success of these models at explaining and predicting health behaviour, research has shown that adults do not always make logical decisions about health behaviour . . .. The errors in logic and reasoning made by cognitively "sophisticated" adults leads to questions about the ability of relatively unsophisticated children to make health-related decisions based on factual information and logic' (Maddux *et al.*, 1986, p. 29).

This statement should not be misunderstood. Many of the models referred to (e.g. Health Belief Model) may provide a useful research framework for understanding how children think and act in terms of health. However, from a motivational standpoint, it appears unlikely that children will be able to think in such abstract and logical terms about health and its possible causes, and also be motivated by such long-term goals. This does not negate valuable educational efforts to inform children about exercise and health, but it does suggest that such efforts, if the goal is to produce informed *and active adults*, should focus on the behavioural skills of adopting and maintaining an active lifestyle, and the experience of positive emotion and reinforcement through health-related exercise. Based on this, perspectives from developmental and health psychology point to the need for a new approach to the education of children in exercise and health.

The discussion so far suggests that more immediate sensations of fun and excitement, as well as competence and mastery, are much more likely to impact on activity patterns for children.

Intrinsic motivation

Intrinsic motivation refers to motivation to do something for its own sake in the absence of external (extrinsic) rewards. This will often involve positive emotion associated with fun, enjoyment and satisfaction. Intrinsically motivated activities have been referred to as 'autotelic' (self-directing) by Csikszentmihalyi (1975), suggesting that intrinsic motivation is linked with feelings of control, competence and self-determination (Deci & Ryan, 1985).

Evidence with children in the sport setting (Wankel & Kreisel, 1985) shows that the intrinsic factors of fun, excitement and skill development are rated as more important than social or extrinsic factors, such as pleasing the coach or winning. Indeed, social psychologists have often provided evidence to show that the use of extrinsic rewards in intrinsically appealing activities can actually reduce intrinsic motivation under some circumstances (Deci & Ryan, 1985; Kassin & Lepper, 1984).

Intrinsic motivation would appear to be a central construct in understanding children's involvement in PA. Csikszentmihalyi (1975) has proposed that an intrinsic state of 'flow' is experienced when the opportunities for challenge are matched by appropriate skills. He defines flow as 'a peculiar dynamic state – the holistic sensation that people feel when they act with total involvement' (p. 36). He goes on to say that one of the main characteristics of flow experiences 'is that they usually are . . . autotelic – that is, people seek flow primarily for itself, not for the incidental extrinsic rewards that may accrue from it' (p. 36). This supports the idea that, where possible, children should be encouraged to participate in PA for the intrinsic value and fun rather than simply to enhance health or fitness, or to gain an extrinsic reward.

Harter (1981*a*) has identified a number of sub-components of intrinsic–extrinsic motivational orientations with children in the classroom. Her measure of these orientations has been adapted for use in physical education and sport settings by Weiss, Bredemeier & Shewchuck (1985) through the development of the 'Motivational Orientation in Sport Scale' (MOSS). The five scales are summarized in Table 13.1, and comprise challenge, curiosity, mastery, judgement and criteria. The first three factors, according to Harter & Connell (1984), can be clustered to form an 'intrinsic mastery motivation' factor, whereas judgement and criteria form an 'autonomous judgement' factor and refers to the 'cognitive–informational' structures of the child. The author's own data with children have shown that high scores on intrinsic mastery motivation correlate positively with sustained periods of high heart rate in 11–12 year-old boys, but not for girls. Conversely, the same aged girls showed a more extrinsic orientation with periods of sustained high heart rates being negatively correlated with autonomous judgement. Discriminant analysis between the groups of active and less active girls showed that the active girls were more motivated by easy rather than challenging tasks, and were more dependent on the judgement and opinion of the teacher in the physical activity settings (Biddle & Armstrong, 1992). Such results require follow-up, but do show one possible explanation for gender differences in activity patterns in children (see also Biddle & Brooke, in press).

Table 13.1. *Intrinsic/extrinsic motivational orientations proposed by Harter (see Harter & Connell, 1984) and modified for sport and physical education by Weiss et al. (1985)*

Motivational orientation	Explanation
Intrinsic mastery motivation	
Challenge	Preference for challenge (intrinsic pole; I) versus preference for easy work (extrinsic pole; E).
Curiosity	Intrinsic interest to satisfy one's own curiosity and interest (I) versus wanting to please the teacher or obtain good grades (E).
Mastery	Preference for independent mastery attempts where the child wants to work out problems and try tasks (I) versus a dependence on the teacher for guidance and help (E).
Autonomous judgement	
Judgement	Feeling capable of making own judgements (I) versus a reliance on the judgement of the teacher about what to do (E).
Criteria	An internal sense of whether one has been successful or unsuccessful (I) versus a dependence on external sources of evaluation, such as grades or teacher feedback (E).

Competence and mastery

Harter's (1978) theory of competence motivation has been suggested as a viable framework for the study of motivation of children in sports and PA (Harter, 1981*b*; Horn & Hasbrook, 1986; Weiss, 1986). White's (1959) notion of 'effectance motivation', where people are thought to be motivated by competence, challenge and curiosity, had a strong influence on Harter's theory. However, the unitary construct of effectance motivation was disputed by Harter (1978) and she has proposed that children hold differentiated perceptions of competence, such as in physical, social, and cognitive domains (Harter, 1982). In addition, Harter (1978) proposed that motivational orientation, as described in the previous section, and perceptions of control, are important constructs within a model of competence motivation for children (see Harter & Connell, 1984).

Essentially, Harter (1981*a*) has predicted that perceived competence, actual competence, motivational orientation, and perceived control interrelate as follows: children who are high in intrinsic mastery motivation are more likely to be those who perceive themselves as competent and feel in control of and responsible for their successes. This is likely to heighten feelings of enjoyment and satisfaction. However, in a study with 250 children aged 6–10 years old, Ulrich (1987) found that perceived physical competence was unrelated to participation in sport, although actual motor competence was related. Ulrich (1987) found that the main reasons the children gave for sport participation were having fun and being with friends. She suggested that these motives are typical of young children, whereas competency-based constructs may become more important with increasing age. Indeed, Weiss, Bredemeier & Shewchuck (1986), in a study of 155 8–12 year-olds, found that higher levels of perceived physical competence correlated with a preference to perform hard and challenging tasks as opposed to easy tasks assigned by the teacher.

Competence motivation and feelings of mastery may be one theoretical avenue to explore in the study of positive affect and children's participation in exercise and PA. However, as yet, the research has focussed only on the competitive sport experience, so little is known about habitual activity levels of children in relation to competence and mastery perceptions.

Children: summary and implications

The case has been made that a priority needs to be placed on understanding the transition from childhood to active adulthood rather than solely on ways of achieving activity/fitness in childhood. A cognitive–developmental framework would suggest that many children are unable to understand the long-term consequences of preventive health actions, and that current motivation and involvement is likely to be related to feelings of intrinsic motivation, control, perceived competence, and enjoyment of the exercise and PA environment. Elsewhere, some basic practical suggestions have been made for motivating children to become active adults (Fox, 1991). First, an emphasis must be placed on the intrinsic joy of PA rather than the extrinsic goals of awards, winning or simply getting fit. Secondly, programmes in physical education may require a major rethink in terms of motivation for health-related exercise (as opposed to sport performance or motor competence). 'Excellence in health is not measurable in terms of running and jumping performance or who wins a game. Health is an outcome of behavioural phenomena which, if it has to be assessed at all, should be judged by how well we maintain appropriate lifestyle patterns' (Fox, 1991). This requires a move away from the 'product' of performance towards the 'process' of participation in order to meet the objective of higher levels of participation that are often stated for improvements in public health. Thirdly, programmes of PA for children and youth should provide varied experiences so that as many children as possible can find a reason to want to be included, and not just those considered to be good performers.

Adults

The majority of the studies on exercise adherence have involved adults in supervised settings. Research has often attempted to discriminate between adherers and non-adherers ('dropouts') on the basis of psychological, socio-demographic and biological factors (for a review see Biddle & Mutrie, 1991). However, adherence is not an 'all-or-none' phenomenon (Sonstroem, 1988), and participation in exercise may involve several different stages and levels of involvement. The model proposed by Sallis & Hovell (1990) provides a useful framework in which to study the factors of exercise adherence. This is illustrated in Figure 13.1 and shows that, in order to understand fully the process of exercise participation, it is necessary to look at the factors that may influence the transitions from sedentary behaviour to adoption, from adoption to maintenance, from adoption to dropout, from maintenance to dropout, and from dropout to resumption. Each may have determinants that are more or less important to that stage compared with other stages (Biddle & Smith, 1991). Indeed, just to illustrate how complex the adherence process could be, it is possible to have at least 210 permutations on determinants! This is calculated on the basis of six developmental periods: early childhood, preadolescence, adolescence, young adulthood, middle age, and geriatric (Maddux *et al.* 1986); seven categories of determinants: beliefs and attitudes, self-perceptions, personality and motivation, social and environmental, biological, mental health outcomes, and self-regulatory skills; these represent very speculative categories! (Biddle & Smith, 1991), and the five stages of adherence given above and identified by Sallis & Hovell (1990).

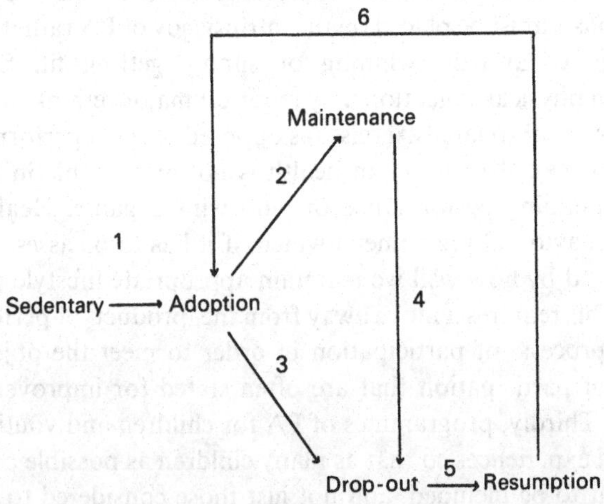

Figure 13.1. An exercise process model. (Adapted from Sallis & Hovell, 1990.)

A selection of key determinants will be discussed for adults using the model proposed by Sallis & Hovell (1990). For the sake of simplicity and space, routes 3 and 4 have been merged, so that the discussion will take place on adoption, maintenance, dropout and resumption.

Adoption of physical activity and exercise

Attitudes

The notion that positive attitudes towards PA and exercise will predict the adoption of an active lifestyle has intuitive appeal. However, the prediction of specific behaviours (e.g. jogging) from generalized attitudes has been unsuccessful. Contemporary attitude theory supports the proposition that a specific behaviour is best predicted by intentions, which, in turn, is related to attitudes and social influences. Perceptions of behavioural control may also influence intention and action. The models illustrated in Figure 13.2 summarize the Theory of Reasoned Action (TRA) and Theory of Planned Behaviour (TPB) (Ajzen, 1985; Fishbein & Ajzen, 1975). Both models have been tested in exercise contexts with moderate success (for reviews see Biddle & Mutrie, 1991; Godin & Shephard, 1990). Research has shown that the attitudinal component of the TRA can predict intentions to exercise, and hence practical intervention at the level of personal beliefs and attitudes may be beneficial. The normative component of the TRA model has been found to be less predictive of exercise behaviours.

When adding the variable of perceived behavioural control to the TRA to form the TPB, Schifter & Ajzen (1985) found that whereas *intentions* to lose weight were related to both attitudinal and normative components, as

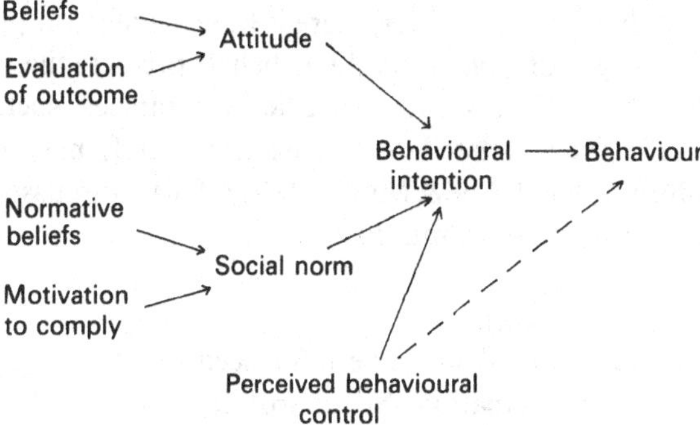

Figure 13.2. The theories of reasoned action and planned behaviour. The diagram shows the complete Theory of Planned Behaviour, while the Theory of Reasoned Action does not include the variable perceived behavioural control.

predicted by the TRA, *actual* weight loss correlated only with perceived control and not with attitudes or social norm factors. Similarly, Gatch & Kendzierski (1990) found that intention to exercise was predicted by perceived control over and above that accounted for by attitudes and social norms.

In summary, attitude measures will predict behaviour more strongly when measures are specific to the behaviour in question. However, other factors will predict intentions and behaviours and these may vary in their impact for different groups. Interventions which attempt to increase exercise adoption should be based on the beliefs, outcome evaluations, and social norm factors of the target population (Smith & Biddle, 1992).

Perceptions of control

The TPB has already shown that perceptions of control may be important in exercise intentions. Indeed, it is often suggested that individuals need to 'take charge' of their lifestyle if behaviour change is to occur. Much of the research in this area has centred on Rotter's (1966) construct of 'locus of control (LOC) of reinforcements', or the extent to which people perceive that reinforcements are within their own control (internal control), controlled by others, or are due to chance (external control). Measures have been of three types: general LOC, health LOC, and exercise/fitness LOC. To date, the evidence is not impressive. Biddle & Mutrie (1991) concluded that 'the often-stated belief that "control" over behaviour is necessary for individuals to lead active or healthy lives has not been supported by the LOC literature. This is probably due to a combination of weak methodology, inadequate instrumentation and the likelihood that LOC is, in any case, only a small part of the explanation of exercise behaviours' (p. 110). Taken in conjunction with other constructs of control, such as attributions and intrinsic motivation, LOC beliefs may be influential to a small degree. However, there is little evidence at present to suggest that the strength of *generalized* LOC beliefs predicts the adoption of exercise or PA. Nevertheless, more specific expectancies, such as self-confidence or self-efficacy related to the exercise itself, may be more predictive (Bandura, 1986; Biddle & Mutrie, 1991). This is discussed later in the section on relapse and resumption.

Barriers to participation

The adoption of exercise is likely to be influenced quite strongly by the perceived barriers to participation. Steinhardt & Dishman (1989), in a psychometric study of perceived benefits and barriers to exercise, found that four major barriers could be identified: time, effort, obstacles, and 'limiting health', although these were not all robust across student and worksite samples. However, Sechrist, Walker & Pender (1987), in a similar study, also found barriers related to time and physical effort, in addition to those of family encouragement and the 'exercise milieu'. These results support the barriers identified in descriptive epidemiological surveys (Canada Fitness Survey, 1983). In short, interventions aimed at exercise adoption need to address the perceptions of barriers, and, in particular, time and effort. This could be achieved through educational strategies aimed at time management and the benefits and opportunities associated with low intensity PA. These barriers are more likely to be associated with exercise rather than habitual PA.

Maintenance of physical activity and exercise

Although the factors affecting adoption and maintenance overlap, the two processes are also likely to involve different factors. In this section, emphasis will be given to the mental health outcomes of participation that are likely to motivate and reinforce participation, and the self-regulatory skills that may be required to maintain activity levels.

Mental health outcomes

The study of the psychology of exercise is characterized by two major themes: the determinants of exercise participation and the mental health outcomes of involvement (Biddle & Mutrie, 1991). However, rarely is the rather obvious point made that the two are likely to be highly interrelated. In other words, positive mental health outcomes will motivate involvement. Despite this, much of the advice given about exercise and health usually centres on the physical benefits, many of which are long term. These are unlikely to be strong motivators and, as Dishman, Sallis & Orenstein (1985) have pointed out, 'feelings related to well-being and enjoyment seem more important to maintaining activity than concerns about health' (p. 166). Exercise psychologists, therefore, as pointed out in the earlier discussion on children, need to know more about the enjoyment of exercise and its function in different groups and individuals.

Morgan & Goldston (1987) summarized the current state of knowledge in exercise and mental health by stating that 'exercise has beneficial emotional effects across all ages and in both sexes' (p. 156). Similarly, Stephens (1988), after an analysis of large-scale community surveys which included data on PA and mental health, concluded that 'the level of physical activity is positively associated with good mental health . . . when mental health is defined as positive mood, general well-being, and relatively infrequent symptoms of anxiety and depression' (p. 41).

The mechanisms for such mental health effects are not well known, although biochemical (Steinberg & Sykes, 1985), physiological (deVries, 1987), and psychological factors have all been implicated. Psychological mechanisms have included changes in self-esteem (Gruber, 1986; Sonstroem, 1984), feelings of satisfaction and mastery (Robbins & Joseph, 1985), and distraction/diversion effects, although studies have not been clear on this factor (Bahrke & Morgan, 1978; Moses *et al.*, 1989). Some have suggested that a feeling of addiction to exercise can be a positive experience (Glasser, 1976), while others say that addiction can lead to serious physical, social and psychological problems (Veale, 1987).

Clearly, the realm of exercise and mental health is important in understanding the motivation of adults in maintaining exercise. However, many questions are unanswered and, in particular, in the application of practical strategies to assist participants in gaining mental health outcomes from PA and exercise.

Self-regulatory skills

The maintenance of activity levels may also require the application of strategies that regulate the individual's behaviour. These 'self-regulatory skills' may be applied easily by some, such as those who are highly self-motivated (Dishman & Gettman, 1980), while others may need assistance in developing these skills over time.

A number of strategies may prove helpful in exercise maintenance (see Biddle & Mutrie, 1991; Knapp, 1988). These may include self-reinforcement, goal-setting, self-monitoring, time management, social support, positive self-statements, and decision balance sheets. Two of the most widely used are goal-setting and self-monitoring.

Goal-setting has been studied in exercise settings (Atkins *et al.*, 1984; Epstein *et al.*, 1980), and positive changes in behaviour are usually found, although long-term follow-up rates are either low or unknown. Similarly, self-monitoring is a cognitive-behavioural technique that has been demonstrated to increase exercise adherence (Martin *et al.*, 1984; Oldridge & Jones, 1983). Examples of self-monitoring include a written diary of progress, and monitoring of heart rate over prolonged exercise periods (Juneau *et al.*, 1987).

In summary, maintenance of exercise can be enhanced by maximizing the feelings of well-being and mental health, and by utilizing self-regulatory strategies.

Dropping out of PA and exercise

Many factors have been identified as potential influences on dropping out of exercise. Mihalik *et al.* (1989) found that such factors changed as a function of the adult life cycle. For example, they found that those aged between 18 and 28 years, and over 50 years, increased their involvement, whereas those aged 29–36 years decreased participation. This is suggestive of social–environmental factors, such as family commitments and grown children leaving home. However, one important factor thought to be associated with dropping out of adult exercise programmes, particularly in supervised settings, is that of exercise intensity.

Exercise intensity

The American College of Sports Medicine (ACSM, 1990), in their revised guidelines on the recommended quality and quantity of exercise for fitness development and maintenance, have stated that cardiorespiratory fitness is best developed in healthy adults when the exercise takes place for no less than 3 days per week for 20–60 minutes at a minimum threshold of intensity of 60% of maximum heart rate (50% of $\dot{V}O_2max$ or heart rate reserve). Lower intensity thresholds are appropriate for those low in fitness. However, these guidelines are often used as the only prescription for exercise in adult exercise classes. While self-regulation of exercise intensity is an important safety factor in such classes, and the use of ratings of perceived exertion (RPE) are commonplace, the effects of such intensities on adherence requires further investigation. Although higher intensities predict greater drop-out, the extent that different individuals prefer different intensities is not well documented (Dishman, 1987; Ingjer & Dahl, 1979). Indeed, research has shown that the anxiety reduction effects of exercise may be related to moderate rather than high intensity exercise (Steptoe & Bolton, 1988; Steptoe & Cox, 1988). Similarly, social psychological research into perceptions of sex-role orientation have shown that higher RPEs have been reported for the same relative workload by those rating themselves as 'feminine-typed' rather than 'masculine-typed' or androgynous (Hochstetler, Rejeski & Best, 1985; Rejeski & Sanford, 1984). Rejeski's (1981) social psychophysiological model of effort perception proposes that exercise tolerance may partly be related to past experience with fatigue, and the social context of the behaviour. Those with limited experience of exercise fatigue or those who are inhibited by the social situation, such as when perceiving the situation to be 'sex-role inappropriate', will inflate effort ratings.

The physiologically driven guidelines for fitness development (ACSM, 1990) require comment in a number of ways. First, it must be recognized that cardiorespiratory fitness is only one fitness outcome from exercise. People may be motivated for gains in strength, flexibility, or reductions in fat, all of which require different recommendations. Second, and as recognized in the ACSM (1990) statement, the guidelines are for fitness, not necessarily for health. The statement says that 'ACSM recognizes the potential health benefits of regular exercise performed more frequently and for a longer duration, but at lower intensities than prescribed in this position statement' (p. 266). Indeed, Dishman (1988*b*) goes further by suggesting that 'the optimal volume of exercise for promoting adherence and health outcomes remains to be identified' (p. 53). This leads to the third point: the relationship between guidelines for exercise/fitness and adherence have all but been ignored. This is illustrated by the ACSM (1990) statement where, after more than four pages of exercise prescription based on physiological principles, the paper concludes with the statement 'emphasis should be placed on factors that result in permanent lifestyle change and encourage a lifetime of physical activity' (p. 270). In some cases, this may require ignoring 'standard' exercise prescriptions if they are too behaviourally challenging!

Relapse and resumption

The literature on adherence to exercise has often assumed that dropping out of supervised exercise equates to a sedentary lifestyle. However, the individual who ceases participation in one setting may continue in another, or start exercising again at a later date. The process of adherence, therefore, is complex and little is known at present about the process of exercise relapse or resumption, labelled the 'start/stop' effect by Biddle & Mutrie (1991).

The concept of behavioural relapse stems from work on addictive behaviours and relapses from drug and alcohol programmes (Brownell *et al.*, 1986; Marlatt, 1985). It has been defined as a 'breakdown or setback in a person's attempt to change or modify any target behaviour' (Marlatt, 1985, p. 3). However, the extent to which addictive behaviours (high frequency, undesired behaviours) can be equated with exercise (low frequency, desired behaviour) remains to be seen.

Two major psychological constructs associated with Marlatt's (1985) relapse prevention model are self-efficacy and attributions.

Self-efficacy

Bandura (1986) has proposed that situation-specific expectancies of efficacy or confidence ('self-efficacy') predict the adoption of behaviour. Bandura (1986) has defined self-efficacy as 'people's judgements of their capabilities to organize and execute courses of action required to attain designated types of performances. It is concerned not with the skills one has but with judgements of what one can do with whatever skills one possesses' (p. 391). To apply this to exercise relapse, Bandura would predict a lower incidence of drop out from exercise programmes for those confident in the beliefs that they can sustain their involvement. Much of the research on self-efficacy and exercise has shown a positive relationship between efficacy beliefs and participation, and this has been for patients in rehabilitation (Ewart, 1989), as well as in community samples (Sallis *et al.*, 1986).

Marlatt (1985) suggests that self-efficacy is related to beliefs that one does or does not possess adequate coping responses in situations where relapse is most likely. Even when exercise may prove difficult and dropping out seems likely, an adequate coping response (e.g. appropriate exercise planning) may increase self-efficacy and decrease the probability of ceasing exercise. Alternatively, an inadequate coping response can lead to decreased feelings of efficacy, a period of inactivity, and possible feelings of guilt and lack of control. These may, in turn, increase the probability of sustained inactivity.

Attributions

The way people appraise events can be important in determining the future reactions towards those events. In the context of exercise, an individual appraising why he/she ceased participation will lead to causal statements or 'attributions'. Attributions can have powerful consequences for emotions, cognitions and future behaviours. In the context of relapse and resumption, it is possible that inactivity will be sustained if people attribute their 'relapse' to factors that they feel are out of their own control (e.g. 'lack of skills in the exercise setting'). However, if they attribute the inactivity to controllable factors, such as poor time management, they are more likely to feel confident about changing the situation. Research has shown large differences in attributions given for adherence and non-adherence to exercise by adults in a supervised setting (Smith & Biddle, 1991).

Conclusions and implications for public health promotion

This chapter has reviewed key areas in the psychology of exercise adherence. Children and adults were dealt with separately as they are likely to involve different processes. For children, the important issue is in establishing positive feelings about PA such that adult involvement is more likely. Using methods appropriate to adult exercise, such as prolonged periods of activity, may be inappropriate, particularly for younger children. They are much more likely to be involved in spontaneous intermittent bouts of PA strongly influenced by fashion and seasonality.

For adults, the process of adherence was discussed in terms of adoption, maintenance, dropping out, and relapse/resumption. As Sallis & Hovell

Table 13.2. *Summary table of key variables thought to influence the routes for adults identified in Figure 13.1*

Variables	Routes			
	1	2	3/4	5
Beliefs and attitudes	**	*	*	*
Self-perceptions:				
Confidence and control	**	**	**	**
Personality and motivation	*	**	**	*
Social and environmental	**	*	**	**
Biological	*	**	***	*
Mental health outcomes	**	***	***	*
Self-regulatory skills	**	***	***	***

* some influence possible.
** expected influence.
*** influence likely to be quite strong.
Note: For simplicity, no distinction is made between moderate levels of physical activity and vigorous exercise, nor to different phases of the adult life cycle, although, as discussed in the chapter, these will probably be factors of some importance.

(1990) point out, little is known about adoption or resumption, since most studies have focussed on the process of maintenance and dropout.

Space has not allowed elaboration on some constructs discussed (see Biddle & Mutrie, 1991, for more detail), and also only a selection of key issues was made. It is likely that some determinants will be very important at one point in the adherence process, and less important elsewhere. An attempt has been made to summarize this in tabular form (see Table 13.2). However, whilst Table 13.2 may have heuristic value, it is very much in need of verification.

Public health campaigns have often relied on messages concerned with the physical effects of exercise and PA, particularly in terms of coronary risk. Similarly, position statements (e.g. Fentem, Bassey & Turnbull, 1988) are heavily biased towards physical outcomes. Important and worthy though these statements are, they fail to address the fundamental issue of how the behaviour is sustained long enough to achieve desired physical and mental benefits. The continued study of the psychological factors of adherence, therefore, is required. While many other factors are possible correlates of exercise participation, such as socio-economic status, biological factors, social and environmental factors, these may be 'distal' antecedents since they are more removed in time from immediate decision making. Psychological constructs may prove to be the 'proximal end points' immediately prior to decision-making (Biddle & Mutrie, 1991).

References

Ajzen, I. (1985). From intentions to actions: a theory of planned behaviour. In *Action Control: From Cognition to Behaviour*. Kuhl, J. and Beckmann, J., eds, Berlin: Springer-Verlag.

American College of Sports Medicine (1990). Position stand: The recommended quantity and quality of exercise for developing and maintaining cardiorespiratory and muscular fitness in healthy adults. *Medicine and Science in Sports and Exercise*, **22**, 265–74.

Armstrong, N., Balding, J., Gentle, P. & Kirby, B. (1990*a*). Patterns of physical activity among 11 to 16 year old British children. *British Medical Journal*, **301**, 203–5.

Armstrong, N., Balding, J., Gentle, P., Williams, J. & Kirby, B. (1990*b*). Peak oxygen uptake and physical activity in 11 to 16 year olds. *Pediatric Exercise Science*, **2**, 349–58.

Atkins, C.J., Kaplan, R.M., Timms, R.M., Reinsch, S. & Lofback, K. (1984). Behavioural exercise programmes in the management of chronic obstructive pulmonary disease. *Journal of Consulting and Clinical Psychology*, **52**, 591–603.

Bahrke, M.S. & Morgan, W.P. (1978). Anxiety reduction following exercise and meditation. *Cognitive Therapy and Research*, **2**, 323–33.

Bandura, A. (1986). *Social Foundations of Thought and Action: A Social Cognitive Theory*. Englewood Cliffs, NJ: Prentice-Hall.

Bibace, R. & Walsh, M. (1979). Developmental stages in children's conceptions of illness. In *Health Psychology*, Stone, G.C. *et al.*, eds, San Francisco: Jossey-Bass.

Biddle, S.J.H. & Armstrong, N. (1992). Children's physical activity: an exploratory study of psychological. *Social Science and Medicine*, **34**, 325–31.

Biddle, S.J.H. & Brooke, R. (1992). Intrinsic versus extrinsic motivational oritentations in phyiscal education and sport. *British Journal of Educational Psychology*, in press.

Biddle, S.J.H. & Fox, K.R. (1989). Exercise and health psychology: emerging relationships. *British Journal of Medical Psychology*, **62**, 205–16.

Biddle, S.J.H., & Smith, R.A. (1991). Motivating adults for physical activity: towards a healthier present. *Journal of Physical Education, Recreation and Dance*, **62**, 39–43.

Biddle, S.J.H. & Mutrie, N. (1991). *Psychology of Physical Activity and Exercise: A Health-Related Perspective*. London: Springer-Verlag.

Bouchard, C., Shephard, R.J., Stephens, T., Sutton, J.R. & McPherson, B.D. (1990) (eds). *Exercise, Fitness and Health: A Consensus of Current Knowledge*. Champaign: Human Kinetics.

Brownell, K.D., Marlatt, G.A., Lichtenstein, E. & Wilson, G.T. (1986). Understanding and preventing relapse. *American Psychologist*, **41**, 765–82.

Canada Fitness Survey (1983). *Fitness and Lifestyle in Canada*. Ottawa: Canada Fitness Survey.

Caspersen, C.J., Powell, K.E. & Christenson, G.M. (1985). Physical activity, exercise and physical fitness: definitions and distinctions for health-related research. *Public Health Reports*, **100**, 126–31.

Csikszentmihalyi, M. (1975). *Beyond Boredom and Anxiety*. San Francisco: Jossey-Bass.

Deci, E.L. & Ryan, R.M. (1985). *Intrinsic Motivation and Self-Determination of Human Behaviour*. New York: Plenum.

deVries, H. (1987). Tension reduction with exercise. In *Exercise and Mental Health*, Morgan, W.P. and Goldston, S.E., eds, Washington, DC: Hemisphere.

Dishman, R.K. (1987). Exercise adherence and habitual physical activity. In *Exercise and Mental Health*, Morgan, W.P. and Goldston, S.E., eds, Washington, DC: Hemisphere.

Dishman, R.K. (1988*a*) (ed.). *Exercise adherence: Its impact on public health*. Champaign: Human Kinetics.

Dishman, R.K. (1988*b*). Behavioural barriers to health-related physical fitness. In *Epidemiology, Behaviour Change, and Intervention in Chronic Disease*, Hall, L.K. and Meyer, G.C., eds, Champaign: Life Enhancement Publications.

Dishman, R.K. & Gettman, L.R. (1980). Psychobiologic influences on exercise adherence. *Journal of Sport Psychology*, **2**, 295–310.

Dishman, R.K., Sallis, J.F. & Orenstein, D.R. (1985). The determinants of physical activity and exercise. *Public Health Reports*, **100**, 158–71.

Engstrom, L-M. (1986). The process of socialisation into keep-fit activities. *Scandinavian Journal of Sports Science*, **8**, 89–97.

Epstein, L.H., Wing, R.R., Thompson, J.K. & Griffin, W. (1980). Attendance and fitness in aerobic exercise. *Behaviour Modification*, **4**, 465–70.

Ewart, C.E. (1989). Psychological effects of resistive weight training: implications for cardiac patients. *Medicine and Science in Sports and Exercise*, **21**, 683–8.

Fentem, P., Bassey, E.J. & Turnbull, N.B. (1988). *The New Case for Exercise*. London: Sports Council and Health Education Authority.

Fishbein, M. & Ajzen, I. (1975). *Belief, Attitude, Intention and Behaviour: An Introduction to Theory and Research*. Reading, Mass.: Addison-Wesley.

Fox, K.R. (1991). Motivating children for physical activity: towards a healthier future. *Journal of Physical Education, Recreation and Dance*, **562**, 34–8.

Gatch, C.L. & Kendzierski, D. (1990). Predicting exercise intentions: the theory of planned behaviour. *Research Quarterly for Exercise and Sport*, **61**, 100–2.

Glasser, W. (1976). *Positive Addiction*. New York: Harper & Row.

Godin, G. & Shephard, R.J. (1990). Use of attitude–behaviour models in exercise promotion. *Sports Medicine*, **10**, 103–21.

Gruber, J.J. (1986). Physical activity and self-esteem development in children: a meta-analysis. In *Effects of Physical Activity in Children*, Stull, G.A. and Eckert, H.M., eds, Champaign: Human Kinetics and American Academy of Physical Education.

Harter, S. (1978). Effectance motivation reconsidered: toward a developmental model. *Human Development*, **21**, 34–64.

Harter, S. (1981*a*). A model of mastery motivation in children: Individual differences and developmental change. In *Minnesota Symposium on Child Psychology, vol. 14*, Collins, A., ed., Hillsdale, NJ: Erlbaum.

Harter, S. (1981*b*). The development of competence motivation in the mastery of cognitive and physical skills: is there still a place for joy? In *Psychology of Motor Behaviour and Sport – 1980*, Roberts, G.C. and Landers, D.M., eds, Champaign: Human Kinetics.

Harter, S. (1982). The perceived competence scale for children. *Child Development*, **53**, 87–97.

Harter, S. & Connell, J.P. (1984). A model of children's achievement and related self-perceptions of competence, control, and motivational orientation. In

Advances in Motivation and Achievement: III. The Development of Achievement Motivation, Nicholls, J., ed., Greenwich, CT: JAI Press.

Hochstetler, S.A., Rejeski, W.J. & Best, W.L. (1985). The influence of sex-role orientation on ratings of perceived exertion. *Sex Roles*, **12**, 825–35.

Horn, T.S. & Hasbrook, C. (1986). Informational components influencing children's perceptions of their physical competence. In *Sport for Children and Youth*, Weiss, M. and Gould, D., eds, Champaign: Human Kinetics.

Ingjer, F. & Dahl, H.A. (1979). Dropouts from an endurance training programme: some histochemical and physiological aspects. *Scandinavian Journal of Sports Science*, **1**, 20–2.

Juneau, M., Rogers, F., DeSantos, V., Yee, M., Evans, A. & Bohn, A. (1987). Effectiveness of self-monitored, home-based moderate intensity exercise training in middle-aged men and women. *American Journal of Cardiology*, **60**, 66–70.

Kassin, S.M. & Lepper, M.R. (1984). Oversufficient and insufficient justification effects: Cognitive and behavioural development. In *Advances in Motivation and Achievement: III. The Development of Achievement Motivation*, Nicholls, J., ed., Greenwich, CT: JAI Press.

Knapp, D.N. (1988). Behavioural management techniques and exercise promotion. In *Exercise Adherence: Its impact on Public Health*, Dishman, R.K., ed., Champaign: Human Kinetics.

Maddux, J.E., Roberts, M.C., Sledden, E.A. & Wright, L. (1986). Developmental issues in child health psychology. *American Psychologist*, **41**, 25–34.

Marlatt, G.A. (1985). Relapse prevention: theoretical rationale and overview of the model. In *Relapse Prevention: Maintenance Strategies in the Treatment of Addictive Behaviours*, Marlatt, G.A. and Gordon, J.R., eds, New York: Guilford Press.

Martin, J.E., Dubbert, P.M., Katell, A.D., Thompson, J.K., Raczynski, J.R. & Lake, M. (1984). Behavioural control of exercise: studies 1 through 6. *Journal of Consulting and Clinical Psychology*, **52**, 795–811.

Mihalik, B.J., O'Leary, J.T., McGuire, F.A. & Dottavio, F.D. (1989). Sports involvement across the lifespan: expansion and contraction of sports activities. *Research Quarterly for Exercise and Sport*, **60**, 396–8.

Montoye, H. (1986). Physical activity, physical fitness, and heart disease risk factors in children. In *Effects of Physical Activity on Children*, Stull, G.A. and Eckert, H.M., eds, Champaign: Human Kinetics and American Academy of Physical Education.

Morgan, W.P. & Goldston, S.E., eds (1987). *Exercise and Mental Health.* Washington, DC: Hemisphere.

Moses, J., Steptoe, A., Mathews, A. & Edwards, S. (1989). The effects of exercise training on mental well-being in the normal population: a controlled trial. *Journal of Psychosomatic Research*, **33**, 47–61.

Natapoff, J.N. (1982). A developmental analysis of children's ideas of health. *Health Education Quarterly*, **9**, 130–41.

Oldridge, N.B. & Jones, N.L. (1983). Improving patient compliance in cardiac rehabilitation: effects of written agreement and self-monitoring. *Journal of Cardiac Rehabilitation*, **3**, 257–62.

Powell, K.E. & Dysinger, W. (1987). Childhood participation in organised school sports and physical education as precursors of adult physical activity. *American Journal of Preventive Medicine*, **3**, 276–81.

Rejeski, W.J. (1981). The perception of exertion: a social psychophysiological integration. *Journal of Sport Psychology*, **3**, 305–20.

Rejeski, W.J. & Sanford, B. (1984). Feminine-typed females: the role of affective schema in the perception of exercise intensity. *Journal of Sport Psychology*, **6**, 197–207.

Robbins, J.M. & Joseph, P. (1985). Experiencing exercise withdrawal: possible consequences of therapeutic and mastery running. *Journal of Sport Psychology*, **7**, 23–39.

Rotter, J.B. (1966). Generalised expectancies for internal versus external control of reinforcement. *Psychological Monographs*, **80**, 1–28.

Sallis, J.F., Haskell, W., Fortmann, S., Vranizan, K., Taylor, C.B. & Solomon, D. (1986). Predictors of adoption and maintenance of physical activity in a community sample. *Preventive Medicine*, **15**, 331–41.

Sallis, J.F. & Hovell, M.F. (1990). Determinants of exercise behaviour. *Exercise and Sport Sciences Reviews*, **18**, 307–30.

Schifter, D.E. & Ajzen, I. (1985). Intention, perceived control and weight loss: an application of the Theory of Planned Behaviour. *Journal of Personality and Social Psychology*, **49**, 843–51.

Sechrist, K.R., Walker, S.N. & Pender, N.J. (1987). Development and psychometric evaluation of the exercise benefits/barriers scale. *Research in Nursing and Health*, **10**, 357–65.

Smith, R.A. & Biddle, S.J.H. (1991). *Exercise adherence in the commercial sector*. In *Proceedings of European Health Psychology Society 4th annual conference*. M. Johnston, M. Herbert and T. Marteau, eds. Leicester: British Psychological Society.

Smith, R.A. & Biddle, S.J.H. (1992). *Attitudes and health-related exercise: Review and critique*. In *Sport and Physical Activity: Moving Towards Excellence*. Williams, T., Almond L. and Sparkes, A., eds, London: E. & F.N. Spon.

Sonstroem, R.J. (1984). Exercise and self-esteem. *Exercise and Sport Sciences Reviews*, **12**, 123–55.

Sonstroem, R.J. (1988). Psychological models. In *Exercise Adherence: Its Impact on Public Health*, Dishman, R.K., ed., Champaign: Human Kinetics.

Steinberg, H. & Sykes, E. (1985). Introduction to symposium on endorphins and behavioural processes: review of literature on endorphins and exercise. *Pharmacology, Biochemistry and Behaviour*, **23**, 857–62.

Steinhardt, M. & Dishman, R.K. (1989). Reliability and validity of expected outcomes and barriers for habitual physical activity. *Journal of Occupational Medicine*, **31**, 536–46.

Stephens, T. (1988). Physical activity and mental health in the United States and Canada: evidence from four population surveys. *Preventive Medicine*, **17**, 35–47.

Stephens, T., Jacobs, D.R. & White, C.C. (1985). A descriptive epidemiology of leisure-time physical activity. *Public Health Reports*, **100**, 147–58.

Steptoe, A. & Bolton, J. (1988). The short-term influence of high and low intensity physical exercise on mood. *Psychology and Health*, **2**, 91–106.

Steptoe, A. & Cox, S. (1988). Acute effects of aerobic exercise on mood. *Health Psychology*, **7**, 329–40.

Ulrich, B.D. (1987). Perceptions of physical competence, motor competence, and participation in organised sport: their interrelationships in young children. *Research Quarterly for Exercise and Sport*, **58**, 57–67.

Veale, D. (1987). Exercise dependence. *British Journal of Addiction*, **82**, 735–40.

Wankel, L.M. & Kreisel, P.S.J. (1985). Factors underlying enjoyment of youth sports: sport and age group comparisons. *Journal of Sport Psychology*, **7**, 51–64.

Weiss, M.R. (1986). A theoretical overview of competence motivation. In *Sport for Children and Youth*, Weiss, M. and Gould, D., eds, Champaign: Human Kinetics.

Weiss, M.R., Bredemeier, B.J. & Shewchuck, R.M. (1985). An intrinsic/extrinsic motivation scale for the youth sport setting: a confirmatory factor analysis. *Journal of Sport Psychology*, **7**, 75–91.

Weiss, M.R., Bredemeier, B.J. & Shewchuck, R.M. (1986). The dynamics of perceived competence, perceived control, and motivational orientation in youth sport. In *Sport for Children and Youth*, Weiss, M. and Gould, D., eds, Champaign: Human Kinetics.

White, R. (1959). Motivation reconsidered: the concept of competence. *Psychological Review*, **66**, 297–333.

14 *Women's working behaviour and maternal–child health in rural Nepal*

CATHERINE PANTER-BRICK

Introduction

There continues to be a great deal of concern for the health of women and young children in rural third-world communities where sustained physical effort is a mandatory way of life. Mothers and children were identified as vulnerable population groups when it was realized that rural women do not necessarily improve their diet or substantially reduce their work-loads during childbearing, and that children of working mothers may experience low birth weight and reduced levels of care. This situation has raised two important areas of inquiry. First, debate has centred on the ways in which women on tight energy budgets manage to cope with repeated reproductive demands (Rajalaksmi, 1980; Prentice, 1984; Durnin, 1985; Durnin & Drummond, 1988; Ferro-Luzzi, 1988). Secondly, the potential conflict between a woman's economic and childcare activities has been highlighted (Leslie, 1988; Hill & Kaplan, 1988), prompting investigation of the cost at which mothers, with little time and energy to spare, combine both responsibilities.

A focus on women's actual behaviour is important to such an inquiry. Whenever possible, women can be expected to minimize their work-loads; in Amazonia for instance, Ache mothers reduce their subsistence activity once children are born, in order to focus on infant care (Hurtado *et al.*, 1985). Yet women may not always be able to curtail their economic activities: they will tend to decrease optional tasks rather than essential economic work. In rural Gambia, women reduced physical activity by 25% in the last term of pregnancy, through curtailing time for housework and leisure, not time for farming (Roberts *et al.*, 1982). It was shown that their work affected birth weight and child mortality, as well as their own nutritional status and reproductive performance (Prentice & Prentice, 1988). An examination of women's activity, their latitude for choice and the impact of their behaviours on maternal and child health, is particularly important for women who assume heavy work-loads and who cannot afford to cut back on economic activities outside the home while childbearing (Panter-Brick, in press).

One is easily convinced that substantial work-loads for women entail repercussions on the health of mothers and young children. However, there is still little information regarding the ways in which women might adopt behaviours to minimize risks to their health, and the extent to which their responses might be successful. How do women cope with a demanding work schedule and maintain well-being? One might expect that women facing real time and energy shortages would organize themselves in energetically efficient ways, in order to sustain obligatory levels of work, assume additional childbearing responsibilities or cope with chronic ill-health.

The inter-relationships between behaviours, health and physical activity are examined for women in rural Nepal, drawing from a year's fieldwork in a Tamang community nestled in the foothills of the Himalaya. In the village of Salme, women's work shows striking caste and seasonal variation, providing an opportunity to examine women's activity and risks to health in different socio-cultural and ecological situations. Tamang women, the focus of this paper, make a substantial contribution to the household's economy (Acharya and Bennett, 1981). Subsistence activity, its impact on maternal energy balance and childcare patterns, and specific behaviours which enable women to sustain their work-loads are examined in turn.

Tamang subsistence activity

The nature and extent of women's activities are described briefly. Living in the foothills of the Himalaya requires extensive mobility on a rugged terrain: women travel up and down steep slopes with gradients of up to 40°, with and without heavy loads. Subsistence activity also requires sustained work effort from day to day: Tamang families draw on all their available labour force, including those women who may suffer chronic ill-health, or who are pregnant and lactating.

The Tamang of Salme (central Nepal) have access to an entire mountainside, 3070 hectares (7586 acres) rising from 1350 m (4429 ft) to 3800 m (12 468 ft), which includes cultivated areas, pasture land and forest. Five main crops (rice, maize and millet, wheat and barley) are grown on terraces at three successive elevations. Households own an average 1.24 hectares (3.06 acres) of land, but plots are small and dispersed at low, medium and high altitudes. This enables families to stagger their harvests in space and time and to achieve food self-sufficiency; however, it requires extensive mobility on the mountain side and a continual input of labour throughout the year. Families also own an average 9.7 head of cattle, sheep and goats, which are moved from terrace to terrace such that dung and urine fertilize the land before it is ploughed.

Levels of activity were quantified through a detailed time-allocation study and measurements of energy expenditure. The impact of motherhood on work-loads was determined for a representative sample of non-childbearing, 6–9 months pregnant, and 1–35 months lactating Tamang women, by following each individual throughout the day and recording her behaviour minute by minute, repeating sample observations in different seasons of the year (for a total 3601 hours of minute-by-minute data on 297 single days). The energy cost of activities was determined by indirect calorimetry, focussing on tasks which occur frequently or require substantial effort (Panter-Brick, 1992). No measurement of maximum aerobic capacity was made to estimate work capacity, but the impact of anaemia on step-test performance was evaluated. Health status was ascertained through a year-long survey of food intakes, regular anthropometric measurements, examination of haematological status and parasitic infection, and records of demographic events.

Women's work

The agricultural calendar is characterized by a seasonal shift in the rhythm of work (as measured by the time women spend in subsistence activity). In the winter, work-loads are moderate as families tend one crop in one area at a time. They plant wheat and barley at high altitude, then harvest rice in low-lying paddy fields, later cut finger-millet and sow maize at mid-altitude, before returning to high-altitude areas for the harvest of wheat and barley. With the advent of monsoon rains, work-loads increase dramatically as families transplant two crops, rice and millet, at low and mid-elevations. The timing of these agricultural tasks cannot tolerate a delay in schedule, as crops planted late miss the best of their growing season, and those left unharvested risk being eaten by cattle penned on neighbouring plots.

The multiple responsibilities inherent to the agro-pastoralist economy generate a shortage of labour among Tamang families, particularly in the monsoon. This problem puts pregnant and lactating women in a particularly difficult situation. How do mothers manage to combine both economic and childcare responsibilities?

The minute-by-minute time allocation study shows that Tamang women strike a different balance of priorities in the winter and the monsoon, in accordance with the demand for female labour (Panter-Brick, 1989). In late winter, third-trimester pregnant and lactating women spent up to a third

Table 14.1. *Seasonal and between-women differences in outdoor work. Time in outdoor subsistence activity (minutes/day)*

	Non-childbearing			Pregnant/lactating			
Women	N	Mean	(SD)	N	Mean	(SD)	Diff. in min.
Late winter season							
Agriculture	19	81.0	(78.0)	15	43.1	(87.8)	37.9*
Husbandry	19	60.6	(60.6)	15	42.7	(62.7)	17.9
Forest-work	19	18.6	(21.4)	15	13.1	(26.7)	5.5
Travel	19	83.4	(44.9)	15	48.4	(42.2)	35.0*
Rest	19	81.2	(62.4)	15	49.6	(49.1)	31.6
Total outdoors	19	324.8	(133.0)	15	196.8	(160.6)	128.0*
Monsoon season							
Agriculture	19	305.9	(169.3)	32	294.9	(144.3)	11.0
Husbandry	19	40.6	(63.6)	32	19.7	(41.3)	20.9
Forest-work	19	0.0	(—)	32	0.4	(2.3)	−0.4
Travel	19	58.3	(32.3)	32	62.4	(28.3)	−4.1
Rest	19	90.1	(49.4)	32	115.0	(67.4)	−24.9
Total outdoors	19	494.9	(180.7)	32	492.4	(174.3)	2.5

N women (means of several observation days per season).
*$P<0.05$, significance level of differences between women.

(39%) less time in outdoor subsistence work, averaging 3.3 hours/day relative to 5.4 hours/day for non-childbearing women. In the monsoon however, all women, whatever their maternal status, assumed strikingly similar work-loads, spending a total 8.2 hours/day outdoors (Table 14.1). Thus, pregnant/lactating women curtail their physical activity in the winter, a time of relative leisure, but not in the monsoon, when households focus on urgent agricultural tasks. This behavioural strategy reflects the degree to which their work is essential to the household economy.

In the monsoon, the Tamang form labour groups exclusively devoted to agricultural activities, which pool resources to complete tasks more rapidly; pregnant and lactating mothers join these groups and manage a remarkable work effort sustained over a number of consecutive days. The day's work is punctuated by frequent pauses (measured minute by minute), amounting to an average 18 minutes per 60 minutes of activity. Rest/work ratios are higher when women work in labour groups than when they work on their own, as people must sustain their effort over a longer working day. A mother will nurse her child while other women simply pause, smoke or eat snacks in the fields, thus minimizing its interference. None the less, nursing prolongs a woman's overall rest time by 10% (Panter-Brick, 1989).

Thus, childcare responsibilities make an impact on the time allocated to rest at the place of work. Lactating women achieved slightly lower work scores than non-childbearing or pregnant women in the monsoon, but differences were not significant.

Travel time and effort

Travel is an essential feature of Tamang lifestyle. Not surprisingly, villagers have devised ways of minimizing the considerable time and energy required for this activity. The Tamang may sleep overnight with their cattle in mobile shelters on the mountain side. The shelters are dismantled and reassembled on different terraces, every 2–3 days on a circuit that averages 19 km over the year. The time-allocation study shows that women can save as much as 110 min/day (1.8 h) by using cattle shelters close to the place of work and avoiding a return journey back home. Consequently, their mean daily travel time is unexpectedly low and constant throughout the year, approximately 1 h/day (just 8% of a 12 h work-day). In addition, the use of cattle shelters avoids having to carry fertilizer to each and every terrace, a task which would certainly be more energetically expensive than that of transporting the shelters from plot to plot. Pregnant and lactating women travel less in the winter, but do not curtail their mobility at other times of the year.

Load carrying up and down mountain slopes is by far the most spectacular and memorable feature of physical activity in Nepal. The Tamang carry loads with regularity, since agro-pastoralist tasks extend in both time and space. The rice harvest demands particular effort, as men and women carry both grain and straw up the mountain side, shouldering 30 kg or more from the low-lying paddy fields (1350–1600 m) (4429–5250 ft) back to the village at 1870 m (6135 ft). Another seemingly arduous task is wood-carrying, which is intensified each year in the spring as families build up their supplies of firewood before the wet season: for a period of two consecutive weeks, men and women, the young and the aged, join labour groups and carry two loads a day every day; these are tightly bound planks (95 cm (37 in) in length and 20 cm (8 in) in width) weighing between 30 kg (66 lb) and 66 kg (145 lb).

Measurements of indirect calorimetry were made to determine the intensity of the work undertaken. Travel was monitored in spring, as people departed from home along four commonly used paths, and returned in the evening with their loads. Pace varied with the nature of the terrain, and was standardized only to the extent that people tend to travel together in groups. Results are expressed in absolute terms (in measured kcal/min) and also, to follow recent international conventions, as a multiple of basal

Table 14.2. *Weight of moderate to heavy loads carried on different inclines in spring. Mean (SD) and range for 30 loads >10 kg*

Slope	Up	Level	Down
N women	2	8	20
Body weight (kg)	46.3 (3.9)	53.2 (4.6)	45.2 (4.0)
Load weight (kg)	30.0 (2.8)	28.5 (9.1)	38.0 (9.5)
range (kg)	28–32	12–41	22–53
% body weight	65%	54%	84%
range (%)	64%–65%	21%–71%	52%–114%

metabolic rate (BMR, predicted from individual weight and height using the FAO/WHO recommended equations; James & Schofield, 1990).

The weights of loads carried in the spring are shown in Table 14.2. Loads carried on level ground averaged 28.5 kg (63 lb) or 54% of body weight, but ranged from 12 kg (26 lb) to 41 kg (90 lb), or 21% to 71% of body weight. Loads carried uphill averaged 30 kg (66 lb) or 65% of a woman's body weight (only two women carried heavy weights uphill when energy cost measurements were made). Loads carried downhill can be very heavy (one woman carried 114% of her own body weight); they averaged 38 kg (84 lb) or 84% of body weight, and ranged widely depending upon whether women carried part of a harvest or pre-stacked firewood. Five of the women carrying moderate to heavy loads were in the last term of their pregnancy.

Table 14.3 shows the energy cost of travel on different inclines, including baseline values for walking without a substantial (<10 kg (22 lb)) load. The energy cost of walking up a slope of 25°–40° with less than 10 kg (5.5 kcal/min) is significantly higher ($P<0.001$) than the cost of walking on level ground (4.7 kcal/min), walking downhill empty-handed is rather infrequent ($N=1$; 2.6 kcal/min). The values are in good agreement with those obtained for highland women in Papua New Guinea (Norgan, Ferro-Luzzi & Durnin, 1974). According to the international classification of activities for a reference 55 kg (121 lb) woman (Durnin & Passmore, 1967), walking uphill is heavy-going, while walking on level ground is moderately demanding and walking downhill light in energy expenditure.

Interestingly, women do not increase their energy expenditure significantly when carrying moderate to heavy loads, as they reduce speed in proportion to the weight and the difficulty of the task (there is always a danger of slipping on wet or stony paths). Thus Tamang women who walked with less than 10 kg on level paths averaged 4.7 kcal/min, while

Table 14.3. *The energetic cost of walking and carrying loads self-paced in kcal/min and as a multiple of predicted BMR (57 measures)*

	Up			Level			Down		
	N	Mean	(SD)	*N*	Mean	(SD)	*N*	Mean	(SD)
kcal/min									
0–9 kg load:	19	5.5	(1.0)	7	4.7	(0.8)	1	2.6	(—)
10–39 kg load:	2	4.8	(0.0)	7	3.8	(0.8)	12	3.1	(0.5)
40–55 kg load:	0	—	—	1	3.9	(—)	8	3.6	(0.6)
Multiple of BMR									
0–9 kg load:	19	6.5	(0.9)	7	5.3	(1.0)	1	2.9	(—)
10–39 kg load:	2	5.6	(0.1)	7	4.3	(0.9)	12	3.6	(0.7)
40–55 kg load:	0	—	—	1	4.0	(—)	8	4.2	(0.6)

Impact of slope ($P<0.008$ with 0–9 kg load; $P<0.0005$ with 10–55 kg load).
ns impact of load weight, ns inter-individual variation.

those carrying 10–39 kg (22–86 lb) loads averaged only 3.8 kcal/min (Figure 14.1). As a result, women's load-carrying activities actually fall within the range of moderately demanding tasks (3.5–5.4 kcal/min/55 kg) rather than being classified as heavy tasks (5.5 kcal/min/55 kg or more); the exception is carrying uphill, which remains heavy-going (5.7 kcal/min/55 kg). The results for self-paced activity are similar to those of Lawrence *et al.* (1985) who showed that many tasks undertaken by rural Gambian women were classified as light to moderate, despite appearing physically demanding.

In addition, journeys are punctuated by frequent pauses at resting metabolic rate, averaging a sixth of travel time or 8–12 minutes per 60 minutes of travel. Experiments have shown that productivity is raised and fatigue prevented if demanding work is interspersed with many small rest periods (Muller, 1953). Durnin and Passmore (1967) also remarked that a hill climber would be less tired after an hour of activity if he had paused 5 minutes after every 10 minutes of walking, than if he had climbed continuously and only then had rested the equivalent time.

The present data cannot demonstrate whether Tamang women achieve enhanced physical or metabolic efficiency for load carrying, as a result of their stout physique and mode of balancing loads on their backs and forehead for vertical ascent (as shown for Kikuyu women who carry up to 20% of their body weight on horizontal paths at no extra cost; Maloiy *et al.*, 1986). However, Waterlow (1990) has noted that mountain people are

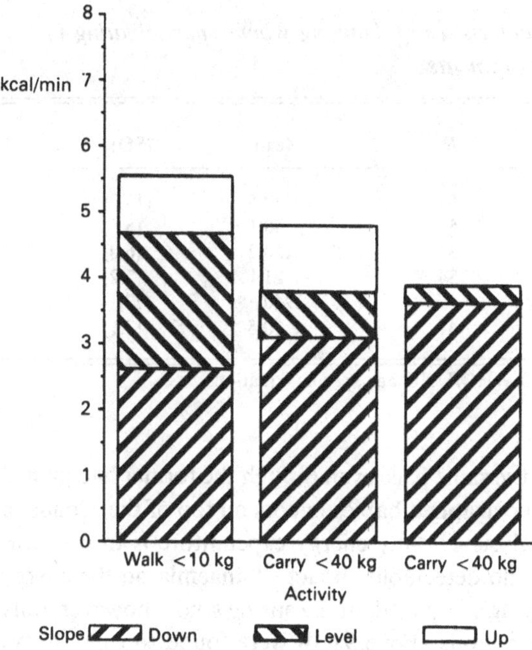

Figure 14.1. Energy cost of walking with and without loads on mountain slopes of different inclines (kcal/min) for Nepali women (the cost of going uphill or on level ground is additional to that of going downhill).

generally small, and that maintaining low body weight is likely to reduce the energy cost of carrying loads up a mountain slope. Tamang women average a mere 48 kg (106 lb) and a stature of 1.5 m (4.9 ft) (Table 14.4), thus carrying loads close to their centre of gravity and to ground level; in addition, carrying loads begins early in childhood, enhancing physical training and habituation to this difficult task. Waterlow (1990) also shows that slow travel speed is likely to reflect a concern for minimizing energy expenditure: over long distances, the slow walker can achieve a greater economy of energy than a fast one.

Additional constraints on work effort

The slow pace of travel, the frequent pauses, and the fact that carrying loads (unless uphill) requires moderate rather than intense effort, may well allow women to sustain their work-loads when facing additional constraints such as childbearing or ill-health. The measurements of energy expenditure for travel (as multiples of BMR) show that pregnant and

Table 14.4. *Sample characteristics of Tamang women participating in indirect calorimetry measurements*

	N	Mean	(SD)
Age (yr)	55	32.5	(13.0)
Weight (kg)	55	48.1	(5.6)
Height (cm)	34	150.2	(6.0)
Body mass index (kg/m^2)	34	21.1	(2.2)
Basal metabolic rate*	55	1240	(77)
Haemoglobin (g/dl)	41	10.5	(2.1)

*kcal/day predicted from FAO/WHO/UNU age specific equations.

lactating women kept to the same pace even though the former had gained body weight and the latter strapped babies in cots on top of their loads (a more comprehensive analysis of daily energy expenditure is in preparation). Moreover, there is no detectable impact of anaemia on the energy cost of walking and carrying self-paced; in the spring study, however, only women with haemoglobin levels above 9 g/dl were found to carry heavy loads.

The impact of anaemia on work performance was tentatively evaluated with a steptest exercise (30 lifts/min). Bench height was varied according to individual body weight, in order to standardize work output and facilitate a within-population comparison with respect to anaemia and energy input. Subjects were divided into two groups according to mean haemoglobin levels (10.94 g/dl for 39 women). Higher heart rates during the third minute of exercise and longer recovery times were found in anaemic men, who worked at 74 watts, but not in the case of anaemic women, who worked at the lower intensity of 55 watts and whose heart rate responses showed considerable overlap (mean (SD) = 126 (15) beats/min). Oxygen uptake during exercise showed no relationship with women's haemoglobin levels. Results suggest that the impact of anaemia on work performance depends on the intensity and the pace of work effort; anaemics can perform their tasks if these are light, or moderately demanding, and if they can work in short bursts followed by periods of recuperative rest (Viteri & Torun, 1974). Moderate anaemia would have a more severe impact at higher work intensities which are sustained without rest.

Studies have shown that anaemia affects maximum aerobic capacity and the capacity to sustain heavy exercise (Viteri & Torun, 1974; Gardner *et al.*, 1977; Collins, 1982; Cotes, Reed & Mortimore, 1982; Haas *et al.*, 1988), but has mitigated effects at light work-loads, e.g. 110 beats/min (Davies,

Chukweumeka, and Van Haaren, 1973; Cotes *et al.*, 1972). Variable effects were also found by Malville (1991), whose sample of Nepali women in the Kathmandu valley (1400 m (4593 ft)) performed a moderate steptest exercise at 40 watts and 18 lifts/min. Below-average haemoglobin levels (< 11.2 g/dl) elevated the heart rates (from 106 to 118 beats/min) and increased the oxygen uptake of Tamang/Newar women, but not of Brahmin/Chetri women who showed a wide range of physiological responses. When considering habitual activity in subsistence tanks, moderate chronic anaemia may not impair absolute work productivity (Collins *et al.*, 1976), particularly if people pace themselves and customarily work at sub-maximal levels.

Health consequences of physical activity

So far attention has been paid to specific aspects of working behaviour, such as the organization of tasks, levels of physical activity, and work pace. The impact of Tamang behaviours on maternal–child health will now be examined, limiting much of the discussion to nutritional considerations. How do Tamang women manage to sustain their habitual levels of physical activity and minimize the risks to their health and to their children? Concern for a woman's energy balance and for a child's well-being arises from the seasonality of Tamang work effort and the extensive mobility of mothers outside the home.

Maternal health and energy balance

An obvious question is to ask how far sustained work effort and seasonal increases of physical activity affect women's nutritional status and reproductive performance. Apart from anaemia, Tamang women show satisfactory nutritional status: they are small but stout, body mass index averaging 21 kg/m^2 for this sample. Mild to moderate anaemia results from hookworm infestation, and iron malabsorption due to excess dietary fibre (Koppert, 1988). Adequate anthropometric status is achieved through securing a steady input of energy from multiple and successive harvests, in return for a sustained work input.

Tamang women lose 2–3% of their body weight during the monsoon, which they then regain in the winter (Koppert, 1988). This small, but significant, weight loss (0.6–1.0 kg (1.3–2.2 lb)) could play a role in determining their low fertility and birth seasonality (completed family size averages 4.7; conceptions fall in the monsoon). Disturbances in women's reproductive function, namely lower ovulation frequency and progesterone levels, have been observed for Lese women in Zaire following a moderate loss of body weight during food shortages, and for American women undertaking moderate or heavy regular exercise (Ellison & Lager, 1986; Ellison, Peacock & Lager, 1989). Ellison (1990) has marshalled convincing evidence that woman's fecundity is responsive to changes in energy balance (loss of body weight and physical activity), not just maternal energy reserves (critical levels of body weight or fat). Nepal provides a good setting for further testing of this hypothesis, in examining the consequences of a seasonal increase in physical activity rather than a seasonal shortage of food. It also allows for the possibility of studying variation in physical activity due to socio-economic lifestyle, by recruiting subjects from different castes in the same geographic environment. The analysis of hormonal profiles from saliva samples is presently being undertaken for two castes in Salme.

Child health and well-being

A second question concerns the impact of a mother's outdoor subsistence activity on childcare patterns. Is child survival compromised by a mother's work away from home and her restricted time for childcare in the monsoon? In the case of the Tamang, childcare is facilitated by the fact that women may bring an infant to the work-place; labour arrangements between households are flexible enough to have nursing mothers join even the most hard-working labour groups. It follows that a child's portability will determine the time he or she spends with the mother, and his or her opportunities for claiming attention.

Tamang mothers carry young infants everywhere, but progressively leave older and heavier children behind. This behaviour has a notable impact on nursing patterns. Mothers breast-feed 1 and 2-year-old infants on demand (for an average 6 minutes at 85 minutes interval throughout the working day); interestingly, nursing times for these young infants show no seasonality despite the increase in maternal work-load during the monsoon (Panter-Brick, 1991). Tamang women seem well able to integrate infant care and outdoor subsistence activity: in the monsoon, high work-loads do not restrict an infant's time on the breast, and furthermore, mothers average similar work scores to those of other women, since they minimize infant interference by nursing during the customary pauses.

However, there is a point at which the strategy no longer works: older children are too cumbersome to carry while still too small to walk. In the monsoon, 3 year-old children are left behind in the village and stay under porches and in alleyways from dawn to dusk, until the adults return home. There is significant seasonality of nursing times for the Tamang three year olds, who are weaned rapidly in the monsoon. Their intake of supplementary food also falls rapidly behind requirements: in Salme, the age-group with the most compromised nutritional status are the 3 to 6 year-old children (Koppert, 1988). Consequently, growth rates are slow, especially in the monsoon. Death rates also rise dramatically at this time of year (by 89% for 1–5 year-olds; Koppert, 1988).

Thus women's outdoor activities affect the health of older children, who are too heavy to carry, more than the health of young infants, who always accompany their mothers (see also Gubhaju, 1985). As discussed elsewhere (Panter-Brick, 1990 and in press) the behaviour of Tamang mothers is none the less an appropriate choice, since mortality rates for children under five are highest in the first year of life (175/1000). The key issue for young infants is one of survival, given the very high death rates before age 1. In contrast, the major risk to the health of older children is one of nutritional well-being, as demonstrated by faltering growth from age 3 to 6.

The ability of Tamang women to integrate economic and child care responsibilities with relative success hinges on their low fertility and long birth intervals, averaging 37.7 months. Death rates for children under five are only 185/1000 where birth spacing is more than 36 months, rising to 296/1000 where birth intervals are short.

Intervening behavioural responses

What general findings can be drawn from this case study? Attention should focus on the particular set of behaviours which mediates the relationship between health and physical capacity. Four socio-cultural strategies are identified below. While these responses are illustrated for Tamang women in Salme, they highlight the role of behaviour as a buffer in the links between health and work capacity (Figure 14.2).

First of all, working behaviour in rural Nepal is characterized by a relatively slow pace, with frequent pauses punctuating short bouts of light to moderately heavy work. Nepali people who travel fairly long distances are accustomed to stopping frequently, for about 10 minutes per hour, at resting places erected at regular intervals along the paths. This behaviour is substantially different to that of western trekkers, who walk relatively speedily but feel exhausted by the end of the day. Frequent rest periods prevent fatigue (Muller, 1953; Durnin & Passmore, 1967), while a slow walking speed can allow for a long-term saving in energy expenditure (Waterlow, 1990). Even physically fit workers cannot sustain efforts greater than 40% of $\dot{V}O_2$max over an 8-hour working day (Spurr, 1990). Alternatively, a slow work pace could itself be a response to poor health and the interference of children. Nepali agricultural labourers alternate

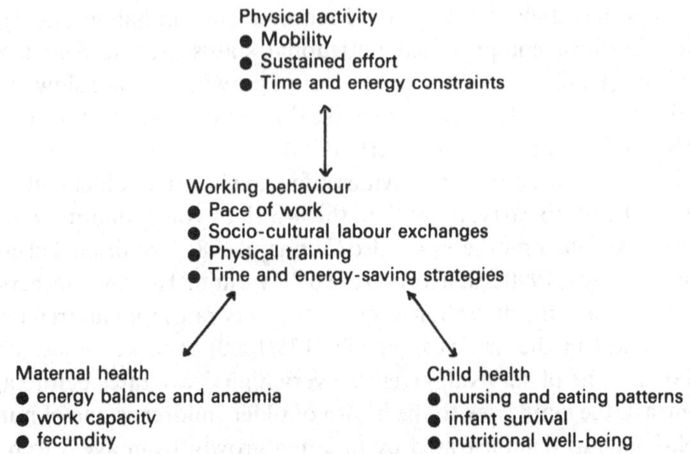

Figure 14.2. Schema representing the mediating role of behaviour on the links between physical activity and maternal-child health.

5–10 minutes of work with short pauses for recuperative rest and also take longer pauses for snacks in the fields. Such frequent breaks may well enable the more anaemic to steady their heart rates or recover oxygen debts, and also allow mothers to nurse on demand. By implication, other workers in a labour group may rest more than they need to (most Nepali stop for a cigarette, but people with severe anaemia or respiratory disease refrain from smoking). In brief, a work schedule with measured tempo and frequent interruptions could well enable women to save energy, steady their heart rates, and nurse their infants at the place of work. This working behaviour appears designed to maximize endurance, rather than productivity: slow pace is essential to sustaining work effort.

Secondly, regular physical training certainly enhances work efficiency. Beall and Goldstein (1988) have demonstrated the importance of lifestyle in determining levels of habitual activity in Nepal: the superior physical fitness of low-caste Sarki men, as measured by heart rate responses to a cycle ergometer test, is explained in terms of their frequent employment as agricultural wage labourers. Similarly, Malville (1991) attributes the enhanced aerobic fitness of Tamang/Newar women relative to the high caste Brahmin/Chetri women in terms of their greater habitual work-loads. The lifestyle determinants of physical activity apply to children as well as to aged men and adult women. Salme children are expected to carry loads at an early age; when 2 or 3 years old, they are given a small bamboo cot or jug of water, weighing 2–4 kg (4.4–8.8 lb) (20–40% of their body weight) to balance on their back and forehead. By age 4 or 5, they may carry a young sibling, and by adolescence they will have assumed the full range of loads carried by the adults. It is hardly an exaggeration to say that young children are taught to carry as soon as they are encouraged to walk; habituation to load-carrying thus occurs throughout their development. Early physical training is a behaviour designed to maximize physical fitness and productivity in adulthood.

Thirdly, there seems to be little discrimination against people who suffer from chronic ill-health or are encumbered by small infants; indeed, these handicaps may be culturally perceived as an inevitable set of circumstances. In Salme, labour exchanges between households, who join forces in the monsoon in order to complete agricultural tasks more rapidly, are reckoned on a per day per person basis, regardless of the age, sex, or physical strength of the individual participants (Panter-Brick, 1989). In more general terms, this situation is more likely to prevail in communities who experience a shortage of labour, rather than a shortage of land and a surplus of hands. Thus in other districts of Nepal, which are more heavily populated, households may refuse to employ women who demonstrate poor productivity (Panter-Brick, unpublished observations). A relatively elastic exchange of labour appears to maximize the use of the available labour force, whereas a more inflexible hiring of labour is geared to achieve maximum productivity.

Finally, specific labour-saving strategies will be adopted to resolve important time and energy constraints. In Salme, these include a careful orchestration of activities in time and space, labour-group arrangements to ensure a rapid completion of tasks, and the use of mobile cattle shelters to reduce the time and energy cost of travel on a rugged terrain (Panter-Brick, 1986 and in press).

Conclusion

Previous research has shown that socio-cultural behaviours constitute an important set of variables influencing the relationship between health, physical activity, work capacity and productivity (Thomas *et al.*, 1988; Beall & Goldstein, 1988). Behavioural responses, which include motivation and skill, can help overcome a demonstrable impairment in work capacity, as shown for Sudanese men who achieved normal levels of productivity despite schistosomiasis infection (Collins *et al.*, 1976). Responses to economic constraints may help explain why hematode infection affects the frequency and length of recuperative rest of women cotton-pickers in the Sudan, but not their overall productivity (Parker, 1989), and why ill-health should have a more detectable impact on one's optional rather than essential tasks. Thomas *et al.* (1988) are currently exploring the biosocial consequences of ill-health for Andean farmers, to examine whether people 'over- or under-compensate behaviourally when confronted with diminished working potential' (p. 269).

This study has focused attention on the situation of working women, whose labour is essential to the household economy, and to a set of behaviours which helps to minimize the impact of outdoor subsistence activity on maternal and child health. The pace of endurance work, the extent of physical training in early childhood, the flexibility of labour exchange and the energetically efficient organization of tasks combine to explain how women manage to sustain demanding levels of physical activity, despite ill-health and a potential conflict between economic and childbearing responsibilities. Behavioural strategies, which mediate the links between health and physical capacity, are as important as mechanical or physiological adaptations, and deserve just as much attention.

Acknowledgements

This study was undertaken in collaboration with the French National Centre of Scientific Research, and financed by The Leverhulme Trust and the Royal Anthropological Society.

References

Acharya, M. & Bennett, L. (1981). *The Rural Women of Nepal: an Aggregate Analysis and Summary of 8 Village Studies*. Kathmandu: Centre of Economic and Development Administration, Tribhuvan University, Nepal.

Beall, C.M. & Goldstein, M.C. (1988). Sociocultural influences on the working capacity of elderly Nepali men. In *Capacity for Work in the Tropics*, Collins, K.J. and Roberts, D.F., eds, pp. 215–26, Cambridge: Cambridge University Press.

Collins, K.J. (1982). Energy expenditure, productivity and endemic disease. In *Energy and Effort*, Harrison, G.A., ed., pp. 65–84, London: Taylor & Francis.

Collins, K.J., Brotherhood, J.R., Davies, C.T.M., Dore, C., Hackett, A.J., Imms, F.J., Musgrove, J., Weiner, J.S., Amin, M.A., El Karim, M., Ismail, H.M., Omer, A.H.S. & Sukkar, M.Y. (1976). Physiological performance and work capacity of Sudanese cane cutters with *Schistosoma mansoni* infection. *American Journal of Tropical Medicine and Hygiene*, **25**, 410–21.

Cotes, J.E., Dabbs, J.M., Elwood, P.C., Hall, A.M., McDonald, A. & Saunders, M.J. (1972). Iron deficiency anaemia; its effects on transfer factor for the lung (diffusing capacity) and ventilation and cardiac frequency during submaximal exercise. *Clinical Science*, **42**, 325–35.

Cotes, J.E., Reed, J.W. & Mortimore, I.L. (1982). Determinants of physical work. In *Energy and Effort*, Harrison, G.A., ed., pp.39–64, London: Taylor & Francis.

Davies, C.T.M., Chukweumeka, A.C. & Van Haaren, J.P.M. (1973). Iron-deficiency anaemia: its effect on maximum aerobic power and responses to exercise in African males aged 17–40 years. *Clinical Science*, **44**, 555–62.

Durnin, J.V.G.A. (1985). Is nutritional status endangered by virtually no extra intake during pregnancy? *Lancet*, **ii**, 823–4.
Durnin, J.V.G.A. & Drummond, S. (1988). The role of working women in a rural environment when nutrition is marginally adequate: problems of assessment. In *Capacity for Work in the Tropics*, Collins, K.J. and Roberts, D.F., eds, pp. 77–83, Cambridge: Cambridge University Press.
Durnin, J.V.G.A. & Passmore, R. (1967). *Energy, Work and Leisure*. London: Heinemann.
Ellison, P.T. (1990). Human ovarian function and reproductive ecology: new hypotheses. *American Anthropologist*, **92**, 933–52.
Ellison, P.T. & Lager, C. (1986). Moderate recreational running is associated with lowered salivary progesterone profiles in women. *American Journal of Obstetrics and Gynecology*, **154**, 1000–3.
Ellison, P.T., Peacock, N.R. & Lager, C. (1989). Ecology and ovarian function among Lese women of the Ituri forest, Zaire. *American Journal of Physical Anthropology*, **78**, 519–26.
Ferro-Luzzi, A. (1988). Marginal energy malnutrition: some speculations on primary energy sparing mechanisms. In *Capacity for Work in the Tropics*, Collins, K.J. and Roberts, D.F., eds, pp. 141–64, Cambridge: Cambridge University Press.
Gardner, G.W., Edgerton, V.R., Senewiratne, B., Barnard, R.J. & Ohira, Y. (1977). Physical work capacity and metabolic stress in subjects with iron deficiency anaemia. *Americal Journal of Clinical Nutrition*, **30**, 910–17.
Gubhaju, B.B. (1985). Regional and socio-economic differentials in infant and child mortality in rural Nepal. *Contributions to Nepalese Studies*, **13** (1), 33–44.
Haas, J.D., Tufts, D.A., Beard, J.L., Roach, R.C. & Spielrogel, H. (1988). Defining anaemia and its effect on physical work capacity at high altitudes in the Bolivian Andes. In *Capacity for Work in the Tropics*, Collins, K.J. and Roberts, D.F., eds, pp. 85–106, Cambridge: Cambridge University Press.
Hill, K. & Kaplan, H. (1988). Tradeoffs in male and female reproductive strategies among the Ache. In *Human Reproductive Behaviour*, Betzig, L., ed., pp. 277–305, Cambridge: Cambridge University Press.
Hurtado, A.M., Hawkes, K., Hill, K. & Kaplan, H. (1985). Female subsistence strategies among the Ache hunter-gatherers of Eastern Paraguay. *Human Ecology*, **13**, 1–28.
James, W.P.T. & Schofield, E.C. (1990). *Human Energy Requirements: A Manual for Planners and Nutritionists*. New York: Oxford University Press (Oxford Medical Publications).
Koppert, G.J.A. (1988). Alimentation et culture chez les Tamang, les Ghale et les Kami du Nepal. These, Faculte de Droit et de Science Politique, Aix-Marseille.
Lawrence, M., Singh, J., Lawrence, F. & Whitehead, R.G. (1985). The energy cost of common daily activities in African women: increased expenditure in pregnancy? *American Journal of Clinical Nutrition*, **42**, 753–63.
Leslie, J. (1988). Women's work and child nutrition in the Third World. *World Development*, **16**, 1341–62.
Maloiy, G.M.O., Heglund, N.C., Prager, L.M., Cavagna, G.A. & Taylor, C.R. (1986). Energetic costs of carrying loads: have African women discovered an economic way? *Nature*, **319**, 668–9.
Malville, N.J. (1991). Haemoglobin concentration, nutritional status, and physical working capacity of nonpregnant rural women of the Kathmandu Valley of Nepal. *American Journal of Human Biology*, **3**, 377–87.

Muller, E.A. (1953). Physiological basis of rest pauses in heavy work. *Quarterly Journal of Experimental Physiology*, **38**, 205–15.

Norgan, N.G., Ferro-Luzzi, A. & Durnin, J.V.G.A. (1974). The energy and nutrient intake and the energy expenditure of 204 New Guinean adults. *Philosophical Transactions of The Royal Society of London, B*, **268**, 309–48.

Panter-Brick, C. (1986). The goths of Salme, Nepal: a strategy for animal husbandry and working behaviour. *Production Pastorale et Société*, **19**, 30–41.

Panter-Brick, C. (1989). Motherhood and subsistence work – the Tamang of rural Nepal. *Human Ecology*, **17**, 205–28.

Panter-Brick, C. (1990). Tamang child care and well-being. *Himalayan Research and Bulletin*, **X**, 1–7.

Panter-Brick, C. (1991). Lactation, birth spacing and maternal work-loads among two castes in rural Nepal. *Journal of Biosocial Science*, **23**, 137–54.

Panter-Brick, C. (in press). Women's work and energetics. In *Women Scientists Look at Evolution: Female Biology and Life History*, Zihlman, A. and Morbeck, M.E., eds.

Panter-Brick, C. (1992). The energy cost of habitual activities in rural Nepal; work pace and sustained work effort. *European Journal of Applied Physiology and Occupation of Physiology*, **64** (in press).

Parker, M. (1989). The effects of *Schistosoma mansoni* on female activity patterns and infant growth in Gezira province, Sudan. D.Phil. thesis, Oxford University.

Prentice, A.M. (1984). Adaptations to long-term low energy intake. In *Energy Intake and Activity*, Pollitt, E. and Amante, P., eds, pp. 3–31, New York: Alan R. Liss.

Prentice, A. & Prentice, A. (1988). Reproduction against the odds. *New Scientist*, **118**, 42–6.

Rajalaksmi, R. (1980). Gestation and lactation performance in relation to the plane of maternal nutrition. In *Maternal Nutrition during Pregnancy and Lactation*, Aebi, H. and Whitehead, R.G., eds, pp. 184–202, Bern: Hans Huber Publishers.

Roberts, S.B., Paul, A.A., Cole, T.J. and Whitehead, R.G. (1982). Seasonal changes in activity, birth weight and lactational performance in rural Gambian women. *Transactions of the Royal Society of Tropical Medicine and Hygiene*, **76**, 668–78.

Spurr, G.B. (1990). The impact of chronic undernutrition on physical work capacity and daily energy expenditure. In *Diet and Disease*, Harrison, G.A. and Waterlow, J.C., eds, pp. 24–61, Cambridge: Cambridge University Press.

Thomas, R.B., Leatherman, T.L., Carey, J.W. & Haas, J.D. (1988). Biosocial consequences of illness among small scale farmers – a research design. In *Capacity for Work in the Tropics*, Collins, K.J. and Roberts, D.F., eds, pp. 249–76, Cambridge: Cambridge University Press.

Viteri, F.E. & Torun, B. (1974). Anaemia and physical work capacity. *Clinical Haematology*, **3**, 609–26.

Waterlow, J.C. (1990). Mechanisms of adaptation to low energy intakes. In *Diet and Disease*, Harrison, G.A. and Waterlow, J.C., eds, pp. 5–23, Cambridge: Cambridge University Press.

15 *Physical activity and psychological well-being*

ANDREW STEPTOE

Introduction

This chapter concerns the effects of physical activity on mood and mental health. There are three major reasons for considering this topic in the context of a symposium on physical activity and health. The first is that mental health is one of the important components of complete well-being that has been associated with exercise. Physical activity may be useful in reducing psychological distress among people in the general population, and in the treatment of more serious problems such as clinical depression and anxiety. Secondly, physical activity may help people cope with stress more effectively, and reduce emotional reactions to stressful life events. Thirdly, an understanding of the beneficial psychological consequences of physical activity may help to enhance adherence to training programmes, and the development of schedules that minimize dropout and foster lifelong active habits. Such information would be valuable both to the enhancement of physical activity in the general population, and to the use of exercise training with special groups such as post-infarction patients (O'Connor *et al.*, 1989).

Evidence linking physical activity with psychological well-being comes from a number of sources. At the anecdotal level, psychological benefits have been trumpeted for many years by exercise enthusiasts. Johnsgård (1989), for example, asserts that he has 'never met a depressed runner'. Comparison of the mood profiles of élite athletes or college sports players with sedentary groups have identified a characteristic profile, with lower scores on tension–anxiety, depression, mental fatigue and confusion scales, and greater mental vigour among the physically active (Morgan & Pollock, 1977; Gondola & Tuckman, 1982). At the epidemiological level, symptoms of depression and lack of psychological well-being have been correlated cross-sectionally with lack of recreational activity in large samples in the USA and Canada (Farmer *et al.*, 1988; Stephens, 1988). Several reviews of physical activity and mental health have been published, and conclusions generally have been rather positive (Dishman, 1985; North, McCullagh &

Tran, 1990; Petruzzello *et al.*, 1991). Indeed, a workshop convened by the US National Institute of Mental Health in 1984 formulated a number of consensus statements to the effect that physical fitness is positively associated with mental health, and that exercise is associated with reductions of emotions such as anxiety and depression (Morgan & Goldston, 1987). However, many of these conclusions have involved confusion between acute and chronic effects, and this field is beset with methodological pitfalls that have only recently been recognized and taken into account. Research designs have become more precise over the last few years, warranting a fresh assessment of the area.

The aim of this chapter is therefore to address the following questions:

1. Does physical activity lead to long-term improvements in mental well-being, after controlling for other influences on psychological state?
2. If so, what type, duration and form of activity is appropriate?
3. What mechanisms are responsible for these responses?

The focus of this chapter is on the long-term effects of training, rather than on the acute after-effects of single bouts of exercise (Morgan, 1985; Steptoe & Bolton, 1988; Steptoe & Cox, 1988). The scope will also be restricted to affective outcomes, rather than other responses to physical training such as changes in cognition or behaviour (Folkins & Sime, 1981; Tomporowski & Ellis, 1986).

Problems in the study of exercise and mental health

There are three general methodological issues that require consideration as a preliminary to discussing empirical links between physical activity and psychological well-being. These concern the appropriate measures of psychological experience, the factors that potentially confound studies of this topic, and the problem of drop-out from activity programmes.

Measures of psychological experience

The first issue is the selection of suitable measures of psychological experience for exercise studies. Many questionnaire and interview schedules are designed to assess psychopathology, and the identification of states of anxiety or depression of a severity that might warrant clinical attention. Although these may be appropriate when the effects of physical activity are assessed in psychiatric populations, applications to the general population may lead to floor effects, with scores being so low before training that the scope for further gains is limited. Measures like the Profile of Mood States developed by McNair and colleagues (1981) are better adapted for assessing affect in the general population, and have been widely used in exercise studies (LeUnas, Hayward & Daiss, 1988). Even these, however, fail to capture the positive feelings of well-being that are more than the mere absence of anxiety, depression or irritation. Watson, Clark & Tellegen (1988) have shown that measures of positive and negative affect are only modestly negatively correlated, reflecting somewhat independent components. Hence, it is sometimes unclear whether studies that show no mood changes with activity programmes document genuine null findings, or whether the measures utilized are insensitive and range restricted.

Several measures of positive well-being have been developed over the last two decades, but these have not been extensively applied in exercise studies (Veit & Ware, 1983; Watson, Clark & Tellegen, 1988; Thayer, 1989; Matthews, Jones & Chamberlain, 1990). Warr (1990) and others have argued that affective well-being can best be understood in relation to the two separate dimensions of hedonic tone (pleasure/displeasure) and arousal. Certain positive feelings are associated with high hedonic tone and high arousal (such as being cheerful and full of energy) while others reflect high hedonic and low arousal (calm, contented). However, Warr has also pointed out the importance of considering 'context-specific' well-being in relation to particular situations or activities. The approach that is used in our research has been to devise measures of positive well-being that are particularly relevant to exercise and physical activity (Moses *et al.*, 1989). An inventory of 36 items related to the feelings that might be engendered by training was developed, and these were scored on the same 5-point scales that are used for the Profile of Mood States. Factor analysis with varimax rotation identified three distinct dimensions that appear to index activity-related well-being. These are shown in Table 15.1. It can be seen that the three dimensions concern positive coping abilities or *coping assets*, perceptions of limitations in coping or *coping deficits*, and perceptions of *physical well-being*. These scales have proved useful in combination with conventional measures of mood and psychopathology in delineating the psychological responses to physical activity.

Potential confounders

From the methodological point of view, one of the major difficulties confronting studies on the psychological effects of exercise is how to distinguish genuine responses from other influences on mood and well-being. Undertaking a training programme or having an active lifestyle are complex phenomena involving a number of factors that can affect psychological state. The most important of these factors are listed in Table 15.2. Only if it can be demonstrated that physical activity has a favourable

Table 15.1. *Activity-related psychological well-being*

Factor 1 *Coping assets*	Factor 2 *Coping deficits*	Factor 3 *Physical well-being*
Self-confident	Easily irritated	Refreshed
Enthusiastic	Disappointed with self	Healthy
Uplifted	Calm (−)	Strong
Proud of self	Drained	Supple
Elated	Easily upset	Fit
Invigorated	Distressed	Well
Coping	Bothered	
Achieving something	Overwhelmed	
Overcoming difficulties	Under too much pressure	
Getting close to goals	Run down	
Competent		
Under control		
Attractive		
Well-organized		

effect over and above these influences, can a positive conclusion be drawn about the direct psychological benefits of exercise. It might be argued that the source of responses does not matter, provided that the outcome is good. This is a short-sighted viewpoint for two reasons. First, the documentation of direct effects is essential if any progress is to be made towards identifying mechanisms. Secondly, public health prescriptions about physical activity must rest on a sound empirical base and accurate understanding of the processes involved. If the positive psychological effects of exercise training are actually due to attention, social involvement or some of the other factors listed in Table 15.2, it may not be necessary to increase physical activity at all. Equally beneficial psychological changes might result from capitalizing on these factors in a completely sedentary population. Attempts to gloss over this possibility may ultimately rebound on the credibility of health promotion programmes.

The first problem that has confounded much research in this field concerns selection. People do not choose to exercise at random, and there may be underlying differences between active and inactive people that are themselves responsible for differences in mental health. This is endorsed by the fact that personality differences have been found between fit and unfit people (Hogan, 1989). This means that the cross-sectional comparisons of active and sedentary people summarized earlier in the chapter provide very weak evidence for the psychological effects of exercise. Longitudinal trials provide the only convincing data. However, longitudinal studies in which

Table 15.2. *Potential confounders and explanations of positive effects*

1. Selection/non-random assignment
2. Attentional effects
3. Expectations of positive effects
4. Progressive achievement in a structured programme
5. Diversion
6. Group and social involvement

participants are not randomized into exercise and control groups are also of rather limited value (e.g. Berger & Owen, 1983, 1988; Lesté & Rust, 1984; Williams & Getty, 1986; Simons & Birkimer, 1988).

The next two factors concern the attention that is paid to clients, and the expectations generated about the training programme. There is ample evidence from the literature on treatment of psychological problems such as anxiety and depression that attention and expectations typically have non-specific positive effects. Indeed, it has been found that the expectations induced prior to therapy are strong predictors of outcome, irrespective of the specific content of the treatment (Lick & Bootzin, 1975). These factors are part of what in drug research would be called the placebo effect, and have given rise to great efforts to develop attention–placebo control conditions in psychological treatment research (Kazdin, 1986). The practical implication for studies of physical activity is that comparisons between exercise and no treatment or waiting list controls cannot rule out competing explanations for positive psychological responses. Comparisons between subjects in various exercise groups who attend different numbers of sessions with trainers are also compromised (e.g. Goldwater & Collis, 1985). Efforts also have to be made to measure expectations concerning different experimental conditions prior to the onset of training, to ensure that these are adequately balanced.

The final set of problems listed in Table 15.2 concerns elements that are involved in much exercise training, but are not exclusive to physical activity programmes. These include the fact that people typically progress through a graded set of activities, for instance, becoming able over time to swim for longer and more efficiently, walk more briskly without getting exhausted, or otherwise show greater endurance. In general terms, physical training involves a structured programme of activity with a series of intermediate goals, the successful accomplishment of which may inculcate a sense of achievement and mastery. Exercise programmes also involve diversion away from typical activities and concerns. For many people, it can be an exhilarating experience simply to have a period set aside from which domestic or occupational demands are banished, for a designated 'private' time. Thirdly, many training studies involve group activities and social involvement. The difficulty for exercise studies is that all these factors may in themselves be psychologically beneficial. One set of treatments for clinical depression, for example, is based on the principle of progressively engaging patients in constructive and rewarding activities (Lewinsohn & Hoberman, 1982). It is possible therefore that a group-based treatment programme which involved acquiring a sedentary skill (such as macramé or piano playing) could lead to improvements in mental health and psychological well-being of a similar order to those produced by vigorous activity.

Of course, it should not be expected that the factors listed in Table 15.2 make no contribution to the favourable psychological impact of physical activity programmes. However, researchers should also try to discover whether physical activity has a unique effect. In terms of experimental designs, this means that randomized controlled trials are essential. Although no treatment controls are helpful, the real test of physical activity comes when a vigorous exercise programme is pitted against an attention–placebo condition matched for expectations, progressive achievement, diversion and group involvement. As will be seen below, few studies fulfil these criteria.

Dropout from training studies

The best designed study in the world would produce inconclusive results if participants failed to complete the programme. The issue of non-adherence to exercise training is discussed elsewhere in this volume. The problem is common in randomized controlled trials because participants may not necessarily be favourably disposed to exercise in the first place. The number of drop-outs is likely to vary with the duration of the study. Typically, in this field of research, 8–12 week training programmes have been studied. A drop-out rate (following commencement of training) of at least 20% is generally found in controlled trials, and in some studies the drop-out has been as high as 40–50% (Ward & Morgan, 1984; Doyne *et al.*, 1987). Indeed, it has been argued that 80–85% is the maximum adherence rate to be expected for structured exercise programmes, even in populations with good facilities and positive attitudes to exercise (Martin *et al.*, 1984).

The crucial problem for controlled trials of psychological responses is not so much that drop-out should be minimized, but that drop-out should not be selective across experimental conditions. If selective drop-out occurs, with, for instance, more people staying in vigorous exercise than in control groups, then the comparability of conditions is compromised. It is also important to study follow-up effects as well as pre-training/post-training differences, so as to evaluate the maintenance of responses beyond the intensive phase of the investigation.

Longitudinal studies of exercise and psychological well-being

Well over 20 randomized controlled longitudinal studies of the effects of exercise training on psychological well-being in adults have now been published, of which some three-quarters have appeared over the last 5 years. Trials have been carried out with a wide range of target populations, including students, volunteers from the general adult population, elderly groups, people with psychiatric or medical disorders, and members of special groups such as the police and military. The primary form of physical activity studied has been aerobic exercise (running, brisk walking, swimming or cycling), although there have been many variations in measures, exercise programmes, research designs and comparison conditions. It is not the intention here to provide an exhaustive description of individual studies, but rather to highlight some general trends in the pattern of results. Meta-analyses of the effects of exercise on depression (North *et al.*, 1990) and anxiety (Petruzzello *et al.*, 1991) have been published, but their conclusions should be treated with caution. North *et al.* mixed studies of long-term change with acute responses to single bouts of exercise in a conceptually confused and inappropriate fashion, while the review by Petruzzello *et al.* omitted many of the recent methodologically sound studies while including several unpublished dissertation reports of doubtful value.

The first conclusion to be drawn from this literature is that virtually no study has shown vigorous aerobic exercise to have a deleterious effect on psychological state. The worst outcome has been no change, and this is itself encouraging. The one exception to this conclusion is the unusual problem of over-training in athletes, which can lead to staleness and disturbances in mood (Morgan *et al.*, 1988). However, this response is probably not relevant to the commoner health-related uses of physical activity training.

Secondly, the majority of randomized studies that have compared aerobic exercise with no-treatment or waiting list conditions have shown greater psychological improvements in the active condition (Folkins, 1976; Jasnoski *et al.*, 1981; Lobitz *et al.*, 1983; Long, 1984; McCann & Holmes, 1984; Martinsen, Medhus & Sandvik, 1985; Fasting & Grønningsaeter, 1986; Doyne *et al.*, 1987; Pavett *et al.*, 1987; Roth & Holmes, 1987; Hannaford, Harrell & Cox, 1988; Mutrie, 1988; Veale & Le Fevre, 1988; Blumenthal *et al.*, 1989; Moses *et al.*, 1989; Norris, Carroll & Cochrane,

1990). These differences have been seen not only in anxiety and depression, but in measures of perceived coping abilities, self-efficacy and general indicators of mental health. However, at least three studies have failed to document differences between aerobic exercise and no treatment conditions (Hughes, Casal & Leon, 1986; King *et al.*, 1989; Lennox, Bedell & Stone, 1990). Since these studies were well conducted, it is appropriate to consider possible explanations for the results.

King *et al.* (1989) randomized 120 healthy middle-aged adults to aerobic conditioning versus assessment only for a 6-month programme, with good measures of adherence and fortnightly recording of mental state. The only differences to emerge were in perceptions of fitness and satisfaction with physical shape and weight, and not in measures of depression, anxiety, well-being or stress. The one unusual feature of this study is that the programme was carried out entirely at home on a relatively intensive schedule, with training five times a week for 45–60 minute sessions. As will be seen later, there is some evidence that the demands of very intensive training schedules may outweigh the psychological benefits of physical activity itself. The study by Hughes *et al.* (1986) was exceptional in involving a cross-over design which may have reduced the contrast between conditions, and in using exercise on a treadmill in the laboratory as the activity manipulation. Such a schedule may have proved tedious over an extended period, militating against elevations in mood. The third study involved the comparison of walking/jogging with volleyball and weight-lifting and with no treatment in healthy middle-aged men and women (Lennox *et al.*, 1990). The drop-out rates were over 30%, and it is possible that the psychological measures were not entirely appropriate for this population.

The source of these differences between studies require further clarification. Notwithstanding these negative findings, the majority of randomized controlled trials demonstrate positive psychological changes with vigorous aerobic exercise. However, it was noted in the previous section that comparisons with no treatment conditions do not provide strong tests of the efficacy of physical activity. Comparisons with active controls are much more illuminating. Unfortunately, far fewer studies have involved active control conditions, so conclusions can only be drawn very cautiously. Nevertheless, at least nine studies have been published in which aerobic exercise produced greater mood change than active control conditions. McCann & Holmes (1984) and Roth & Holmes (1987) randomized mildly depressed students to aerobic exercise, relaxation training or no treatment control conditions. Yoga and flexibility exercises were used as the active control in a study of healthy people aged 60–83 (Blumenthal *et al.*, 1989), while a supportive study group was devised by Fasting & Grønningsaeter

(1986) for their investigation of the long-term unemployed. The study reported by Carney *et al.* (1987) was particularly interesting since it involved adults with renal disease on haemodialysis, and compared physical exercise with a support group in which patients were able to discuss their concerns and problems. After six months, exercise had led to a reduction in depression (as indexed by the Beck Depression Inventory), while depression had increased to some extent in the support group. Two other studies with active controls were reported by our research group (Moses *et al.*, 1989; Steptoe *et al.*, 1989). The second of these is described briefly below, in order to provide a concrete illustration of the procedures and pattern of results in this type of investigation.

An illustrative study

The study was designed to assess the impact of moderate intensity aerobic training on the psychological state of sedentary anxious adults from the general population. Respondents to an advertisement asking for volunteers for a study of fitness and health were screened with two measures, and were deemed eligible if they had anxiety ratings in the high normal or borderline clinical range. Participants were randomized into moderate aerobic training or attention–placebo conditions for the 10-week study, with one supervised group session and three unsupervised sessions per week. Moderate aerobic training involved 20 minutes continuous walking or jogging at an intensity of 60–65% of maximal heart rate (HR max). The attention–placebo condition consisted of strength, mobility and flexibility exercises, performed discontinuously for at least 20 minutes at an intensity that did not elevate heart rate above 50% HR maximum. Participants in both conditions were provided with detailed training manuals that outlined the rationale for the programmes. Each course was structured so that individuals began at an activity level appropriate for their fitness, and progressed towards more demanding levels over the programme. Careful attention was paid in each group to helping participants integrate training into their daily routines. Intermediate and long-term goals were set in terms of improvements in endurance in the moderate exercise condition, and flexibility and suppleness in the attention–placebo condition. Participants were taught to monitor their own heart rates in order to ensure that they remained within prescribed limits.

Expectations concerning the likely benefits of participation were assessed with five items, asking the extent to which subjects believed the programme would be enjoyable and improve their level of fitness, sense of well-being and health. There was no significant difference between groups in expectancy ratings, which were generally high. Likewise, ratings of satisfaction taken after training again showed no differences, indicating that the conditions were comparable in terms of enjoyability. It would therefore appear that the criteria listed in Table 15.2 for matching physical activity with active control conditions in terms of attention, expectations, progressive achievement, diversion and social involvement were generally fulfilled.

The number of volunteers beginning the study was 24 in the moderate exercise and 23 in the attention–placebo condition. Seven dropped out of each group during the training period, so the overall adherence rate was 70.2%. Nine of the drop-outs were for good reasons such as moving house or hospital admission. This left a sample of 14 women and three men in the moderate exercise and 14 women and two men in the attention–placebo condition. Adherence to training was also gauged through analysing weekly diaries. Interestingly, these showed that the mean number of sessions per week was significantly higher in the aerobic exercise condition (3.45 versus 3.13). On the other hand, the mean session duration was significantly longer in the attention–placebo condition (44.8 versus 33.0 minutes), so that the lower number of sessions was compensated by longer training.

Figure 15.1 summarizes representative results, with maximum oxygen consumption ($\dot{V}O_2$max) estimated from a sub-maximal bicycle ergometer test, tension–anxiety from the profile of mood states, and coping assets from the measure shown in Table 15.1. It can be seen that aerobic training led to significant improvements in fitness, coupled with greater reductions in tension and increases in perceived coping ability. Greater reductions in depression and mental fatigue and increases in mental vigour were also

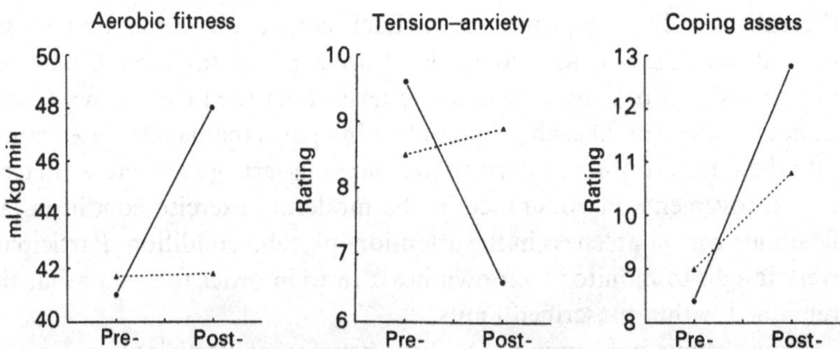

Figure 15.1. Average $\dot{V}O_2$max estimated by cycle ergometry (*left panel*), tension-anxiety from the Profile of Mood States (*centre panel*), and perceived coping assets (*right panel*), recorded before (pre) and after (post) 10 weeks of moderate aerobic exercise (solid line, filled circles) or attention–placebo training (broken line, filled triangles). (Based on Steptoe *et al.*, 1989)

recorded with aerobic training. Some increase in perceived coping ability was reported from participants in the attention–placebo group, reflecting the non-specific effects of that condition. The pattern of results was maintained at three-month follow-up, with significant differences in trait anxiety, tension, depression, mental vigour and coping assets.

This, and other studies, appear therefore to confirm that aerobic training does have a positive effect on psychological well-being over and above that attributable to non-specific factors. However, some good studies have shown no differences between aerobic exercise and active controls. It is notable that the control conditions in these studies have included cognitive therapy, stress inoculation and anxiety management training (Lobitz *et al.*, 1983; Long, 1984, 1985; Fremont & Craighead, 1987; Long & Haney, 1988). These are recognized treatments utilized clinically for people with stress-related problems, and it is therefore not proper to consider them as inert control conditions. Rather, they should be regarded as active comparison treatments. The fact that aerobic training produced equivalent effects is therefore quite an encouraging finding.

Psychological benefits and improvements in fitness

It is possible to conclude on the basis of evidence currently available that aerobic exercise does have a positive effect on psychological well-being for many people. The question then arises of whether the exercise needs to be aerobic, or whether other forms of activity will do. In addition, does the exercise have to be sufficiently vigorous and regular as to improve cardiorespiratory fitness and endurance? The relationship between psychological benefits and fitness is also important for understanding the mechanisms underlying responses. Improved cardiorespiratory fitness is the main consequence of aerobic exercise training, so it is natural to consider whether it is also responsible for psychological effects.

Despite the appealing simplicity of this notion, there are three reasons why improved cardiorespiratory conditon does not appear to mediate psychological gains. The first reservation stems from the effects of anaerobic training on psychological state. In direct comparisons with aerobic exercise, anaerobic training has been shown to produce similar psychological responses in the majority of randomized controlled trials (Doyne *et al.*, 1987; Emery & Blumenthal, 1988; Martinsen, Hoffart & Solberg, 1989*a*,*b*). These results have emerged despite the substantially greater improvements in cardiorespiratory fitness recorded with aerobic exercise. Perhaps the best examples of this pattern are two studies carried out in Norway involving psychiatric in-patients with primary diagnoses of depression and anxiety (Martinsen *et al.*, 1989*a*,*b*). In both studies, aerobic training was carried out over an eight-week period, with three one-hour sessions per week at 70% $\dot{V}O_2$max. Patients in non-aerobic groups carried out muscular strength and flexibility exercises at a low intensity on the same schedule. In each case, significant increases in aerobic capacity were only recorded in aerobic training groups. Yet the same changes in depression and anxiety were found in aerobic and anaerobic conditions. Two other trials have shown slight advantages for aerobic exercise over anaerobic programmes, but only in the short-term (Mutrie, 1988; Norris *et al.*, 1990). Of course, it can be argued that these studies do not conclusively rule out the mediating role of cardiorespiratory fitness, since different mechanisms might be operative in anaerobic and aerobic programmes. This is not, however, a very parsimonious explanation.

Secondly, a number of studies have failed to show any significant correlation between changes in fitness and psychological responses within aerobic training groups (Fasting & Grønningsaeter, 1986; Hughes *et al.*, 1986; Fremont & Craighead, 1987; Emery & Blumenthal, 1988; Steptoe *et al.*, 1989). Admittedly, measurement error in estimating fitness change may militate against reliable associations. Yet it seems clear that although aerobic exercise leads to greater elevations in mood than do control conditions, within aerobically trained groups the subjects who become fittest are not necessarily the ones whose psychological well-being is most improved. This links with the third argument, which is that no systematic dose–response association between training intensity and psychological response has been recorded. Several studies have compared aerobic exercise conducted at intensive and moderate levels (McPherson *et al.*, 1967; Sexton, Maere & Dahl, 1989; Stevenson & Topp, 1990). Similar psychological improvements were recorded at both exercise intensities. Indeed, in some circumstances, intensive programmes may have a deleterious effect. This was seen in a study from our research group, in which moderate aerobic training (60–65% HR max.) was compared with more intensive training (75–80% HR max.) and an attention–placebo condition in sedentary adults from the general population (Moses *et al.*, 1989). Figure 15.2 summarizes some of the changes induced by the ten-week programme. Cardiorespiratory fitness responses were proportional to the training stimulus, with the largest improvement in the intense condition and the least change in the attention–placebo condition. However, psychological state improved to the greatest extent in the moderate training group, with significant differences in tension–anxiety and mental confusion. It was hypothesized that the benefits of exercise were offset in the intensive training group by the high demands on this sedentary population.

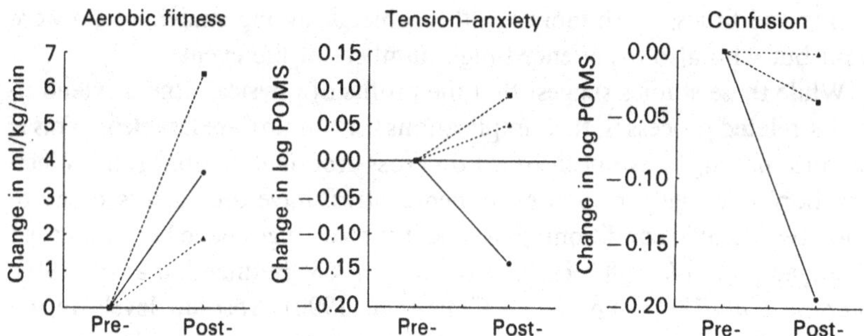

Figure 15.2. Change in average $\dot{V}O_2$max estimated from the 12-min walk–run test (*left panel*), tension–anxiety (*centre panel*), and mental confusion (*right panel*) from the Profile of Mood States, recorded before (pre) and after (post) 10 weeks of high intensity aerobic exercise (broken line, filled squares), moderate aerobic exercise (solid line, filled circles) or attention–placebo training (broken line, filled triangles). (Based on Moses *et al.*, 1989)

Fitness and resistance to psychological stress

The literature therefore suggests that no simple relationship exists between improved cardiorespiratory fitness and the long-term psychological effects of exercise. However, physical fitness may be an important factor moderating the impact of mental and emotional demands on stress-related physiological responses and illness vulnerability. This aspect of stress tolerance is rather different from the components of mental health that have been discussed thus far, since it is concerned with the psychophysiological consequences of adverse life experience, rather than affective responses. Stress tolerance can be modelled in the laboratory by comparing psychophysiological responses in fit and unfit groups. Such experiments typically involve measurement of heart rate, blood pressure, electrodermal activity, catecholamines or other stress-related physiological parameters, while subjects are challenged by demanding problem-solving tasks or other stimuli. Although results have been inconsistent, a number of investigators have reported lower autonomic and neuroendocrine reactivity in fitter subjects, or else more rapid recovery following mental stress tests (for reviews, see Crews & Landers, 1987; Steptoe, 1989). There is also evidence from ecological studies that physical fitness may moderate the influence of life stress on health and well-being (Pavett *et al.*, 1987; Brown & Siegel, 1988; Roth *et al.*, 1989). For example, Brown (1991) assessed cardiorespiratory fitness, life events and mood in a group of students, and recorded illness-related visits to the university health centre as an indicator of health outcome. Health centre visits were predicted by the interaction between life events and fitness, with more health problems among students who were unfit but who also experienced high numbers of life events.

While these studies suggest that the profits of physical fitness extend to stress-related processes, their implications for training are less clear. This is because findings have been based on cross-sectional comparisons. When psychophysiological responses to mental stress have been assessed before and after aerobic conditioning, frequently no changes have been observed (Seraganian *et al.*, 1987; Holmes & Roth, 1988; Blumenthal *et al.*, 1990; Steptoe *et al.*, 1990; in press; De Geus *et al.*, 1990). Absolute levels during mental stress tests of parameters such as heart rate may be reduced following training, but reactivity has typically remained constant. Similarly, ecological studies have demonstrated the stress-buffering effects of fitness rather than physical activity level (Roth *et al.*, 1989; Brown, 1991). It may be that constitutional differences between fit and unfit people are responsible for the stress-buffering effects, or else that training needs to be sustained for longer than the typical 8–12 weeks before effects emerge. Cardiorespiratory fitness and vigorous physical activity may have separable but overlapping influences on different aspects of mental health.

Mechanisms underlying psychological responses

The evidence described in the last section suggests that improvements in cardiorespiratory fitness are unlikely to account for the positive effects of physical activity on mental well-being. On the other hand, it seems that a certain level of vigorous (aerobic or anaerobic) activity is required, since very low level programmes are not efficacious. What processes are then responsible? Two possibilities have been investigated, and both may play some part.

After-effects of individual exercise sessions

A session of vigorous exercise has several short-term physiological and psychological after-effects. The first possibility is that long-term psychological changes derive from the accumulation of acute responses. For example, there may be a reduction in anxiety or inhibition of stress responsivity immediately following an exercise session. As people maintain regular activity schedules, they may spend an increasing proportion of waking life in a more positive mood state. This might, in turn, modify their cognitive appraisal of problems that arise on an everyday basis, leading to sustained psychological benefit.

The evidence related to this hypothesis is inconclusive at present. Positive mood changes in the period immediately following exercise sessions have been reported by a number of investigators (see Morgan, 1985). However, it has previously been argued that many of these studies have been biased by the expectations of researchers and subjects about the favourable effects of exercise. When the purpose of the study is disguised, or mood is treated as an incidental rather than a key measure, the results are far less positive (Steptoe & Bolton, 1988; Steptoe & Cox, 1988). One popular theory is that exercise-induced increases in beta-endorphin are responsible for short-term elevations in mood (Allen & Coen, 1987). Here again, the evidence is inconsistent, since a number of studies have shown no differences in psychological responses to exercise in double-blind comparisons of naloxone and placebo (Markoff, Ryan & Young, 1982; Farrell *et al.*, 1986; McMurray *et al.*, 1988). Correlations between increases in beta-endorphin and mood improvement have even been negative in some cases (Kraemer *et al.*, 1990).

The possible role of this acute mechanism has been investigated by assessing psychophysiological responses to mental stress tests, performed a short time after exercise sessions at different workloads. Roy & Steptoe (1991) compared 20 minutes of exercise at 100W, 25W, and a no-exercise control, in a study of healthy young men. All subjects carried out a stressful mental arithmetic test 20 minutes after exercise or control periods. The result was a clear reduction in heart rate, systolic and diastolic blood pressure responses to mental arithmetic among subjects who had worked at 100W, compared with controls. The reactions of the 25W group were intermediate. Acute inhibition of physiological stress responsivity following vigorous exercise was therefore apparent. A second, more elaborate study has produced somewhat similar findings, although differences were less striking. Unfortunately, neither of these studies throws as much light as might have been hoped on the issue. Although cardiovascular activity was reduced, psychological responses (self-reported mood) were unaffected by prior exercise. Hence it is possible that the diminution in cardiovascular responsivity to mental stressors is a purely physiological effect, resulting from the hypotensive state that typically follows vigorous exercise (Kaufman, Hughson & Schaman, 1987).

The role of acute changes in psychological and psychophysiological state has probably not yet been assessed in an appropriate fashion. All the studies to have been conducted thus far have taken a snapshot, single session approach to the issue. Short-term responses may evolve and accumulate over time, in which case it may be necessary to follow subjects longitudinally, assessing changes in the intensity and course of exercise after-effects.

Exercise-related cognitions

A second possible mechanism underlying improvement in psychological well-being with physical training concerns the generalization of changes in exercise-related cognitions. Sonstroem & Morgan (1989) have argued that an important early result of participating in exercise training is an improvement in physical self-efficacy, or the confidence people have in their ability to achieve the specific tasks involved in the programme. With practice, people may feel increasingly confident in being able to jog 1 mile, walk for 1 hour, or whatever the training involves. Success at the series of graded mastery experiences that constitute a progressive training programme may, in turn, enhance beliefs in physical competence, and more general feelings about the abilities and capacities of one's body. Body satisfaction may likewise increase. These changes in perceptions of physical competence and body satisfaction may in turn enhance self-esteem and self-concept, leading to benefits at the affective level.

This model makes clear suggestions about the way specific experiences with physical activity may generalize to broader areas of psychological function. Similar notions have been put forward by others (Ewart, 1989; Ossip-Klein *et al.*, 1989). There is substantial support for the notion that self-esteem is an important determinant of affective state, and links have been established with depression (Brown, Bifulco & Andrews, 1990). Several studies have shown that self-concept and general measures of self-efficacy and confidence are augmented by physical training (Long & Haney, 1988; Ossip-Klein *et al.*, 1989). These responses might, however, simply be components of the mental health outcome of training, rather than part of the mediating process. As yet, the only direct test of this model has been in a study of late middle-aged adults (Sonstroem *et al.*, 1991). Although it was cross-sectional, structural modelling with LISREL was used to tease out the associations between variables. Sonstroem *et al.* found that the best fit model was consistent with the hypothesis, in that specific exercise-related self-efficacies predicted physical competence which in turn predicted self-esteem. Confirmation of these patterns in longitudinal studies is now needed.

If generalization of exercise-related cognitions is the important mechanism, two conclusions follow. Firstly, careful attention must be paid to the exact format and demands of the exercise programme, to ensure that appropriate graded mastery experiences are provided. As indicated earlier (and shown in Figure 15.2), some programmes may be excessively demanding, perhaps setting their aspirations too high. The second implication is that the mechanism underlying psychological responses is not exclusive to exercise. Other types of experience may be equally efficacious, provided they are structured in a suitable way. They might have a parallel impact on affective state through enhancing other aspects of self-efficacy. Increased physical activity may be just one of several processes converging on a common pathway.

Conclusions

Three questions were posed at the beginning of this chapter about the effects of physical activity on well-being. In relation to the first, there now appears to be sufficient evidence to conclude that exercise does have positive long-term effects on mood and psychological well-being, and that these are not accounted for by selection, group socialization, expectations or attentional factors. Favourable psychological responses have been recorded in randomized controlled trials from a wide variety of adult groups. Whether or not such patterns will generalize to the population at large is not known, because the mere fact of volunteering for an exercise trial itself involves some form of self-selection. Even though psychological gains are not restricted to athletes, they may be confined to people with a positive disposition towards increasing their physical activity level.

The second question concerns the forms of physical activity that produce positive mood changes. The evidence currently available suggests that both aerobic and anaerobic exercise are suitable. The intensity of activity required is less clear. It appears that very light activity programmes are not sufficient, and there are also indications that intensive conditioning programmes may confer less benefit than moderately vigorous schedules. The value of moderate intensity exercise has important implications for health promotion and disease prevention. Most of the health benefits of exercise are not immediate, but involve hypothetical reductions in risk of future ischaemic heart disease and disorders of ageing (Powell *et al.*, 1987; Cooper, Barker & Wickham, 1988). Psychological responses are more immediate, and therefore may be crucial in encouraging exercise maintenance. The fact that activity schedules do not have to be very intensive for psychological responses to emerge, and that mood changes do not depend on improved cardiorespiratory fitness, may encourage members of the public to increase activity levels. They may be reassured that training does not have to be taken to rigorous extremes before any health benefit is apparent. Recent literature on the physical health consequences of activity has also emphasized the protective effects of moderate exercise (LaPorte *et al.*, 1984; Haskell, 1985; Leon *et al.*, 1987).

The final question, concerning the mechanisms underlying psychological responses, is among the most interesting issues for current research. In the absence of clear associations with changes in cardiorespiratory or other forms of fitness, alternative mechanisms need to be considered. The two possibilities outlined here concern the generalization of changes in exercise-related cognitions, and the acute after-effects of individual exercise periods. There is preliminary evidence in favour of these processes, and both are promising avenues for future work. The clarification of mediating processes is a key step towards harnessing the potential of vigorous physical activity in the maintenance of health.

References

Allen, M.E. & Coen, D. (1987). Naloxone blocking of running-induced mood changes. *Annals of Sports Medicine*, **3**, 190–5.

Berger, B.G. & Owen, D.R. (1983). Mood alteration with swimming – swimmers really do 'feel better'. *Psychosomatic Medicine*, **45**, 425–35.

Berger, B.G. & Owen, D.R. (1988). Stress reduction and mood enhancement in four exercise modes: swimming, body conditioning, Hatha yoga and fencing. *Research Quarterly in Exercise and Sport*, **59**, 148–59.

Blumenthal, J.A., Emery, C.F., Madden, D.J., George, L.K., Coleman, R.E., Riddle, M.W., McKee, D.C., Reasoner, J.A. & Williams, R.B. (1989). Cardiovascular and behavioral effects of aerobic exercise training in healthy older men and women. *Journal of Gerontology*, **44**, M147–57.

Blumenthal, J.A., Fredrikson, M., Kuhn, C.M., Ulmer, R.L., Walsh Riddle, M. & Appelbaum, M. (1990). Aerobic exercise reduces levels of cardiovascular and sympathoadrenal responses to mental stress in subjects without prior evidence of myocardial infarction. *American Journal of Cardiology*, **65**, 93–8.

Brown, J.D. (1991). Staying fit and staying well: physical fitness as a moderator of life stress. *Journal of Personality and Social Psychology*, **60**, 555–61.

Brown, G.W., Bifulco, A. & Andrews, B. (1990). Self-esteem and depression: 3. Etiological issues. *Social Psychiatry and Psychiatric Epidemiology*, **25**, 235–43.

Brown, J.D. & Siegel, J.M. (1988). Exercise as a buffer of life stress: a prospective study of adolescent health. *Health Psychology*, **7**, 341–53.

Carney, R.M., Templeton, B., Hong, B.A., Harter, H.R., Hagberg, J.M., Schechtman, K.B. & Goldberg, A.P. (1987). Exercise training reduces depression and increases the performance of pleasant activities in hemodialysis patients. *Nephron*, **47**, 194–8.

Cooper, C., Barker, D.J.P. & Wickham, C. (1988). Physical activity, muscle strength and calcium intake in fracture of the proximal femur in Britain. *British Medical Journal*, **297**, 1443–6.

Crews, D.J. & Landers, D.M. (1987). A meta-analytic review of aerobic fitness and reactivity to psychosocial stressors. *Medicine and Science in Sports and Exercise*, **19**, S114–20.

De Geus, E.J.C., van Doornen, L.J.P., de Visser, D.C. & Orlebeke, J.F. (1990). Existing and training-induced differences in aerobic fitness: their relationship to physiological response patterns during different types of stress. *Psychophysiology*, **27**, 457–80.

Dishman, R.K. (1985). Medical psychology in exercise and sport. *Medical Clinics of North America*, **69**, 123–43.

Doyne, E.J., Ossip-Klein, D.J., Bowman, E.D., Osborn, K.M., McDougall-

Wilson, I.B. & Neimeyer, R.A. (1987). Running versus weight lifting in the treatment of depression. *Journal of Consulting and Clinical Psychology*, **55**, 748–54.

Emery, C.F. & Blumenthal, J.A. (1988). Effects of exercise training on psychological functioning in healthy type A men. *Psychology and Health*, **2**, 367–79.

Ewart, C.K. (1989). Psychological effects of resistive weight training: implications for cardiac patients. *Medicine and Science in Sports and Exercise*, **21**, 683–8.

Farmer, M.E., Locke, B.Z., Moscicki, E.K., Dannenberg, A.L., Larson, D.B. & Radloff, L.S. (1988). Physical activity and depressive symptoms: the NHANES 1 epidemiologic follow-up study. *American Journal of Epidemiology*, **128**, 1340–51.

Farrell, P.A., Gustafson, A.B., Garthwaite, T.L., Kalkoff, R.K., Cowley, A.W. & Morgan, W.P. (1986). Influence of endogenous opioids on the response of selected hormones to exercise in humans. *Journal of Applied Physiology*, **61**, 1051–7.

Fasting, K. & Grønningsaeter, H. (1986). Unemployment, trait anxiety and physical exercise. *Scandinavian Journal of Sports Sciences*, **8**, 99–103.

Folkins, C.H. (1976). Effects of physical training on mood. *Journal of Clinical Psychology*, **32**, 385–9.

Folkins, C.H. & Sime, W.E. (1981). Physical fitness training and mental health. *American Psychologist*, **36**, 373–89.

Fremont, J & Craighead, L.W. (1987). Aerobic exercise and cognitive therapy in the treatment of dysphoric moods. *Cognitive Therapy and Research*, **11**, 241–51.

Goldwater, B.C. & Collis, M.L. (1985). Psychologic effects of cardiovascular conditioning: a controlled experiment. *Psychosomatic Medicine*, **47**, 174–81.

Gondola, J. & Tuckman, B. (1982). Psychological mood states in 'average' marathon runners. *Perceptual and Motor Skills*, **55**, 1295–300.

Hannaford, C.P., Harrell, E.H. & Cox, K. (1988). Psychophysiological effects of a running program on depression and anxiety in a psychiatric population. *Psychological Record*, **38**, 37–48.

Haskell, W.L. (1985). Physical activity and health: need to define the required stimulus. *American Journal of Cardiology*, **55**, 4D–9D.

Hogan, J. (1989). Personality correlates of physical fitness. *Journal of Personality and Social Psychology*, **56**, 284–8.

Holmes, D.S. & Roth, D.L. (1988). Effects of aerobic training and relaxation training on cardiovascular activity during psychological stress. *Journal of Psychosomatic Research*, **32**, 469–74.

Hughes, J.R., Casal, D.C. & Leon, A.S. (1986). Psychological effects of exercise: a randomized cross-over trial. *Journal of Psychosomatic Research*, **30**, 355–60.

Jasnoski, M.I., Holmes, D.S., Solomon, S. & Aguiar, C. (1981). Exercise, changes in aerobic capacity, and changes in self-perceptions: an experimental investigation. *Journal of Research in Personality*, **15**, 460–6.

Johnsgård, K.W. (1989). *The Exercise Prescription for Depression and Anxiety*. New York: Plenum Press.

Kaufman, F.L., Hughson, R.L. & Schaman, J.P. (1987). Effect of exercise on recovery of blood pressure in normotensive and hypertensive subjects. *Medicine & Science in Sports and Exercise*, **19**, 17–20.

Kazdin, A.E. (1986). Research designs and methodology. In *Handbook of Psychotherapy and Behavior Change*, 3rd edn, Garfield, S.L. and Bergin, A.E., eds, pp. 23–68. New York: John Wiley.

King, A.C., Taylor, C.B., Haskell, W.L. & DeBusk, R.F. (1989). The influence of regular aerobic exercise on psychological health: a randomised, controlled trial of healthy middle-aged adults. *Health Psychology*, **8**, 305–24.

Kraemer, R.R., Dzewaltowski, D.A., Blair, M.S., Rinehardt, K.F. & Castracane, V.D. (1990). Mood alteration from treadmill running – its relationship to beta-endorphin, corticotrophin and growth hormone. *Journal of Sports Medicine and Physical Fitness*, **31**, 241–6.

LaPorte, R.E., Adams, L.L., Savage, D.D., Brenes, G., Deerwater, S. & Cook, T. (1984). The spectrum of physical activity, cardiovascular disease and health: an epidemiologic perspective. *American Journal of Epidemiology*, **120**, 507–17.

Lennox, S.S., Bedell, J.R. & Stone, A.A. (1990). The effect of exercise on normal mood. *Journal of Psychosomatic Research*, **34**, 629–36.

Leon, A.S., Connett, J., Jacobs, D.R. & Rauramaa, R. (1987). Leisure-time physical activity levels and risk of coronary heart disease and death. *Journal of the American Medicine Association*, **258**, 2388–95.

Leste, A. & Rust, J. (1984). The effects of dance on anxiety. *Perceptual and Motor Skills*, **58**, 767–72.

LeUnas, A.L., Hayward, S.A. & Daiss, S. (1988). Annotated bibliography on the Profile of Mood States in sport, 1975–88. *Journal of Sport Behavior*, **11**, 213–39.

Lewinsohn, P.M. & Hoberman, H.M. (1982). Depression. In *International Handbook of Behavior Modification and Therapy*, Bellack, A.S., Hersen, M. and Kazdin, A.E., eds, New York: Plenum Press.

Lick, J. & Bootzin, R. (1975). Expectancy factors in the treatment of fear: methodological and theoretical issues. *Psychological Bulletin*, **82**, 917–31.

Lobitz, W.C., Brammell, H.L., Stoll, S. & Niccoli, A. (1983). Physical exercise and anxiety management training for cardiac stress management in a nonpatient population. *Journal of Cardiac Rehabilitation*, **3**, 683–8.

Long, B.C. (1984). Aerobic conditioning and stress inoculation: a comparison of stress-management interventions. *Cognitive Therapy and Research*, **8**, 517–42.

Long, B.C. (1985). Stress management interventions: a 15-month follow-up of aerobic conditioning and stress inoculation training. *Cognitive Therapy and Research*, **9**, 471–8.

Long, B.C. & Haney, C.H. (1988). Coping strategies for working women: aerobic exercise and relaxation interventions. *Behavior Therapy*, **19**, 75–83.

McCann, I.L. & Holmes, D.S. (1984). Influence of aerobic exercise on depression. *Journal of Personality and Social Psychology*, **46**, 1142–7.

McMurray, R.G., Berry, M.J., Hardy, C.J. & Sheps, D.S. (1988). Physiologic and psychologic responses to a low dose of naloxone administered during prolonged running. *Annals of Sports Medicine*, **4**, 21–5.

McNair, D.M., Lorr, N. & Droppleman, L.F. (1981). *Manual for the Profile of Mood States*. San Diego: Education and Industrial Testing Service.

McPherson, B.D., Paivio, A., Yuhasz, M.S., Rechnitzer, P.A., Pickard, H.A. & Lefcoe, N.M. (1967). Psychological effects of an exercise program for post-infarct and normal adult men. *Journal of Sports Medicine*, **7**, 95–102.

Markoff, R.A., Ryan, P. & Young, T. (1982). Endorphin and mood changes in long-distance running. *Medicine and Science in Sports and Exercise*, **14**, 11–15.

Martin, J.E., Dubbert, P.M., Catell, A.D., Thompson, J.K., Raczynski, J.R., Lake, M., Smith, P.O., Webster, J.S., Sikora, T. & Cohen, R.E. (1984). Behavioral control of exercise in sedentary adults: Studies 1 through 6. *Journal of Consulting and Clinical Psychology*, **52**, 795–811.

Martinsen, E.W., Hoffart, A. & Solberg, O.Y. (1989*a*). Aerobic and non-aerobic forms of exercise in the treatment of anxiety disorders. *Stress Medicine*, **5**, 115–20.

Martinsen, E.W., Hoffart, A. & Solberg, O.Y. (1989*b*). Comparing aerobic and non-aerobic forms of exercise in the treatment of clinical depression: a randomised trial. *Comprehensive Psychiatry*, **30**, 324–33.

Martinsen, E.W., Medhus, A. & Sandvik, L. (1985). Effect of aerobic exercise on depression: a controlled trial. *British Medical Journal*, **291**, 109.

Matthews, G., Jones, D.M. & Chamberlain, A.G. (1990). Refining the measurement of mood: the UWIST Mood Adjective Checklist. *British Journal of Psychology*, **81**, 17–42.

Morgan, W.P., Costill, D.L., Flynn, M.G., Raglin, J.S. & O'Connor, P.J. (1988). Mood disturbance following increased training in swimmers. *Medicine and Science in Sports and Exercise*, **20**, 408–14.

Morgan, W.P. & Pollock, M.L. (1977). Psychologic characterisation of the elite distance runner. *Annals of the New York Academy of Science*, **301**, 382–403.

Morgan, W.P. (1985). Affective beneficence of vigorous physical activity. *Medicine and Science in Sports and Exercise*, **17**, 94–100.

Morgan, W.P. & Goldston, S.E. (1987) Summary. In *Exercise and Mental Health*, Morgan, W.P. and Goldston, S.E., eds, pp. 155–60. Washington: Hemisphere.

Moses, J., Steptoe, A., Mathews, A. & Edwards, S. (1989). The effects of exercise training on the mental well-being in the normal population: a controlled trial. *Journal of Psychosomatic Research*, **32**, 47–61.

Mutrie, N. (1988). Exercise as a treatment for moderate depression in the UK Health Service. In *Sport, Health, Psychology and Exercise Symposium*, pp. 96–105. London: Sports Council.

Norris, R., Carroll, D. & Cochrane, R. (1990). The effects of aerobic and anaerobic training on fitness, blood pressure, and psychological stress and well-being. *Journal of Psychosomatic Research*, **34**, 367–75.

North, T.C., McCullagh, P. & Tran, V.Z. (1990). Effect of exercise on depression. *Exercise and Sport Science Review*, **18**, 379–415.

O'Connor, G.T., Buring, J.E., Yusuf, S., Goldhaver, S.L., Olmstead, E.M., Paffenbarger, R.S. & Hennekens, C.H. (1989). An overview of randomised trials of rehabilitation with exercise after myocardial infarction. *Circulation*, **80**, 234–44.

Ossip-Klein, D.J., Doyne, E.J., Bowman, E.D., Osborn, K.M., McDougall-Wilson, I.B. & Neimeyer, R.A. (1989). Effects of running or weight lifting on self-concept in clinically depressed women. *Journal of Consulting and Clinical Psychology*, **57**, 158–61.

Pavett, C.M., Butler, M., Marcinik, E.J. & Hodgdon, J.A. (1987). Exercise as buffer against organisational stress. *Stress Medicine*, **3**, 87–92.

Petruzzello, S.J., Landers, D.M., Hatfield, B.D., Kubitz, K.A. & Salazar, W. (1991). A meta-analysis on the anxiety-reducing effects of acute and chronic exercise. *Sports Medicine*, **11**, 143–82.

Powell, K.E., Thompson, P.D., Caspersen, C.J. & Kendrick, J.S. (1987). Physical activity and the incidence of coronary heart disease. *Annual Review of Public Health*, **8**, 253–87.

Roth, D.L. & Holmes, D.S. (1987). Influence of aerobic exercise training and relaxation training on physical and psychological health following stressful life events. *Psychosomatic Medicine*, **49**, 355–65.

Roth, D.L., Wiebe, D.J., Fillingim, R.B. & Shay, K.A. (1989). Life events, fitness, hardiness, and health: a simultaneous analysis of proposed stress-resistance factors. *Journal of Personality and Social Psychology*, **57**, 136–42.

Roy, M. & Steptoe, A. (1991). The inhibition of cardiovascular responses to mental stress following aerobic exercise. *Psychophysiology*. **28**, 689–700.

Seraganian, P., Roskies, E., Hanley, J.A., Oseasohn, R. & Collu, R. (1987). Failure to alter psychophysiological reactivity in Type A men with physical exercise or stress management programs. *Psychology and Health*, **1**, 195–213.

Sexton, H., Maere, A. & Dahl, N.H. (1989). Exercise intensity and reduction in neurotic symptoms. A controlled follow-up study. *Acta Psychiatrica Scandinavica*, **80**, 231–5.

Simons, C.W. & Birkimer, J.C. (1988). An exploration of factors predicting the effects of aerobic conditioning on mood state. *Journal of Psychosomatic Research*, **32**, 63–75.

Sonstroem, R.J. & Morgan, W.P. (1989). Exercise and self-esteem: rationale and model. *Medicine and Science in Sports and Exercise*, **21**, 329–37.

Sonstroem, R.J., Harlow, L.L., Gemma, L.M. & Osborne, S. (1991). Test of structural relationships within a proposed exercise and self-esteem model. *Journal of Personality Assessment*, **56**, 348–64.

Stephens, T. (1988). Physical activity and mental health in the United States and Canada: evidence from four population surveys. *Preventive Medicine*, **17**, 35–47.

Steptoe, A. (1989). Psychophysiological interventions in behavioural medicine. In *Handbook of Clinical Psychophysiology*, Turpin, G., ed., pp. 215–39. Chichester: John Wiley.

Steptoe, A. & Bolton, J. (1988). The short-term influence of high and low intensity physical exercise on mood. *Psychology and Health*, **2**, 91–106.

Steptoe, A. & Cox, S. (1988). Acute effects of aerobic exercise on mood. *Health Psychology*, **7**, 329–40.

Steptoe, A., Edwards, S., Moses, J. & Mathews, A. (1989). The effects of exercise training on mood and perceived coping ability in anxious adults from the general population. *Journal of Psychosomatic Research*, **33**, 537–47.

Steptoe, A., Moses, J., Edwards, S. & Mathews, A. (in press). Exercise and responsivity to mental stress: Discrepancies between the subjective and physiological effects of aerobic training. *International Journal of Sport Psychology*.

Steptoe, A., Moses, J., Mathews, A. & Edwards, S. (1990). Aerobic fitness, physical activity and psychophysiological reactions to mental tasks. *Psychophysiology*, **27**, 264–74.

Stevenson, J.S. & Topp, R. (1990). Effects of moderate and low intensity long-term exercise by older adults. *Research in Nursing and Health*, **13**, 209–18.

Thayer, R.E. (1989). *The Biopsychology of Mood and Arousal*. Oxford: Oxford University Press.

Tomporowski, P.D. & Ellis, N.D. (1986). Effects of exercise on cognitive processes: a review. *Psychological Bulletin*, **99**, 338–46.

Veale, D. & Le Fevre, K. (1988). Aerobic exercise in the adjunctive treatment of depression: a randomised controlled trial. In *Sport, Health, Psychology and Exercise Symposium*, pp. 106–11. London: Sports Council.

Veit, C.T. & Ware, J.E. (1983). The structure of psychological distress and well-being in general populations. *Journal of Consulting and Clinical Psychology*, **51**, 730–42.

Ward, A. & Morgan, W.P. (1984). Adherence patterns of healthy men and women enrolled in an adult exercise program. *Journal of Cardiac Rehabilitation*, **4**, 143–52.

Warr, P.B. (1990). The measurement of well-being and other aspects of mental health. *Journal of Occupational Psychology*, **63**, 193–210.

Watson, D., Clark, L.A. & Tellegen, A. (1988). Development and validation of brief measure of positive and negative effect. The PANAS Scales. *Journal of Personality and Social Psychology*, **54**, 1063–70.

Williams, J. & Getty, D. (1986). Effect of level of exercise on psychological mood state, physical fitness, and plasma beta-endorphin. *Perceptual and Motor Skills*, **63**, 1099–105.

16 *Leisure lifestyles: present and future*

SUE GLYPTIS

Leisure plays a significant role in shaping everyday lives and in defining the quality of life. This paper examines the nature and meaning of leisure, outlines recent trends and present patterns of leisure behaviour, and discusses the part played by leisure in shaping present and future lifestyles.

Definitions

Leisure and lifestyle are neither easy to define nor straightforward to measure. Definitions alone have generated a substantial literature (for example de Grazia, 1962; Brightbill, 1963; Dumazedier, 1967; Countryside Recreation Research Advisory Group, 1970; Neulinger, 1974; Kelvin, 1979; Kelly, 1982; Tokarski Filipcova & Glyptis, 1990). Leisure has been defined in three broad ways. The first relates to time. Leisure is frequently referred to as 'the time available to the individual when the disciplines of work, sleep and other basic needs have been met' (Countryside Recreation Research Advisory Group, 1970, p. 5). However, this notion of leisure as a mere leftover denies it any positive character or purpose. In reality, leisure is not simply time free from other things; as many unemployed people and retired people know, free time can be far from leisurely: in large quantity it can be more of a burden than a blessing.

The second approach views leisure more positively as particular types of activity, which may provide rest and recuperation, diversion, excitement, personal and social fulfilment, and, literally, mental or physical re-creation.

The third approach combines elements of the first two, but adds an important subjective perspective. Kelvin (1979), for example, distinguished between 'objective' and 'subjective' free time: a person may be objectively free every Sunday but might feel subjectively bound to do the gardening, visit an elderly relative, or take the children swimming. Leisure, in other words, derives from the meaning of the activity to the individual, and not from the activity itself: an activity which is toil to one person may bring positive pleasure to another, gardening and DIY being obvious examples. Wherever possible, this chapter adopts this subjective approach, taking leisure to mean any activity (or inactivity) freely undertaken for the sake of enjoyment.

Leisure, then, remains mainly an activity-centred concept, we speak of spending our leisure for badminton, boating, bee-keeping or bingo. But any activity in isolation reveals very little about the meaning of leisure to individual people, whether the people who take part in one sporting activity are more or less likely to take part in other sports, or whether sports participants are more or less likely to be hobbyists, country lovers or concert goers. Yet it seems likely that people's leisure is developed around a few key interests or influences: as Roberts (1978, p. 37) stated, 'individuals do not so much engage in *ad hoc* miscellanies of activities as develop broader systems of leisure behaviour consisting of a number of interdependent elements'. These 'broader systems', and the motivations which underlie them, are referred to as leisure lifestyles. In this chapter, lifestyle will be taken as the total range of activities undertaken by a given individual.

Leisure activities: recent trends and present patterns

In the heady optimism of the 1960s and early 1970s, there were confident predictions that people were fast moving to an age of leisure. Indeed, if time, money and mobility are the facilitators of leisure there are plenty of measures to show that society is becoming more leisured. Between 1961 and 1988 the normal basic working hours of employees in full time manual occupations fell by almost four hours a week from 42.8 to 38.9 hours, and actual hours worked fell from 45.5 to 43.5 hours. Holiday entitlement has risen substantially: in 1961 only 3% of manual workers were entitled to more than two weeks' holiday; by 1988 all but 1% had four weeks or more (Social Trends, 1990). In addition, with trends towards earlier retirement, and increased life expectancy, there is the prospect of a substantial block of largely free time at the end of a working life. Men reaching the age of 60 in 1990 could expect a further 16.8 years and women a further 21.2 years; those reaching 70 could expect a further 10.5 and 13.7 years respectively.

The increase in free time has been paralleled by an increase in wealth. Real household disposable income per head rose by 40% between 1976 and 1988. Expenditure on many leisure-related items increased dramatically: spending on television and video rose by 142%; spending on vehicle purchase by 109%; and that on vehicle maintenance by 44%. Car ownership has more than doubled since 1961: in 1961 31% of British households had a car; in 1987 64% had at least one car and 19% had two or more (Social Trends, 1990).

Against such a backcloth, all bodes well for leisure participation. This is

Table 16.1. *Sports participation in the United Kingdom*

		% taking part in past month		
		Outdoor		Indoor
		excluding walking	including walking	
Men	1977	23	35	31
	1980	24	37	32
	1986	27	40	35
Women	1977	8	21	13
	1980	9	24	15
	1986	10	24	21

Source: *General Household Survey*, London: HMSO.

borne out by basic leisure participation statistics, though the popularity of different forms of leisure varies. Sport has been a major beneficiary of improved material circumstances and increased public awareness of the benefits of an active lifestyle. Periodic General Household Surveys since 1977 have shown a steady rise in participation rates by the adult population, with the sharpest increase occurring in indoor sports participation by women (Table 16.1). Growth has been particularly marked in individual, as opposed to team based, sports, in pursuits linked with exercise and health, and in the more adventurous and glamorous activities. Membership of the Ramblers' Association, for example, increased from 35 731 in 1980 to 57 936 in 1985; that of the Royal Yachting Association from 65 180 to 73 520; membership of the British Sub Aqua Club, having doubled in the 1970s, increased again from 27 075 to 33 988; and membership of the British Orienteering Federation rose from 22 000 to 50 700 (Countryside Recreation Research Advisory Group, 1980; Sports Council, 1986).

Though its growth has been remarkable, sport overall, however, is still a minority activity. In Britain only 11 sports attract as many as 3% of the population. Of the population, 3% amounts to 1.7 million people, a considerable volume of activity, and substantial spending of time and money. None the less, compared to many other forms of leisure, sport is a poor relation. The Sports Council's ambition to achieve 'Sport for All' remains an ambition, not an accomplishment. The leisure activities with mass appeal are still predominantly home based, passive, informal and social. Virtually everyone (98% of the population) watches television, 94%

Table 16.2. *Sports participation and social class in the United Kingdom*

	% taking part in past month, 1987	
	Professional	Unskilled manual
Swimming	21	5
Football	6	3
Golf	10	1
Cycling	12	6
Tennis	4	1
Squash	9	1
Badminton	7	1
Keep fit, yoga	9	3

Source: *General Household Survey*, 1987, London: HMSO.

visit or entertain friends and relatives, 86% listen to the radio, 67% listen to records and tapes, 59% read books, 55% go to the pub, and 47% go out for meals.

Participation in sport and physical recreation is not only relatively low, but also socially stratified. The social contrasts are diminishing over time, but remain clearly evident. Leisure participation generally, and sports participation in particular, are dominated by men, young people, white people, car owners, and those in white collar occupations. Women, older people, ethnic minorities, non car owners, and people in blue collar occupations have much lower participation rates. These contrasts are evident in a simple comparison between opposite ends of the social class spectrum (Table 16.2): among the professional class, participation in virtually all sports is at least double that of unskilled manual workers, and in some cases as much as ten times greater.

The response of the policy-makers

Public sector leisure providers generally, and the Sports Council especially, have identified a range of 'target groups': sectors of the population whose participation in recreational activities is markedly lower than average, and whose access to opportunities is diminished by a variety of constraints (Sports Council, 1988). These groups are well documented. They include the retired, whose health, mobility and financial resources may be declining, who may be lacking in companions or confidence. Many have been neither accustomed to having free time on their hands, nor introduced to satisfying leisure opportunities in their younger years (Boothby *et al.*,

1981; Long & Wimbush, 1985); they may be, as Rodgers termed them 'sports-illiterate' (Rodgers, 1977). Women are another low-participant group. For many, domestic responsibilities constrain the availability and use of free time, and gender stereotypes inhibit the expression of personal, out-of-home leisure interests (Talbot, 1979; Green, Hebron & Woodward, 1987). Unemployed people are a third such group. They usually have an abundance of free time but few means to enjoy it, and the psychological challenge of structuring unstructured days and finding a real sense of purpose outside the context of work (Kelvin & Jarrett, 1985; Fryer & Ullah, 1987; Glyptis, 1989). Ethnic minorities are also under-represented in sport and recreation. They are over-represented among the poor, the poorly housed and the unemployed, perhaps isolated by barriers of language and custom, and torn between maintaining their own religious and cultural traditions and assimilating into the way of life of their host communities (Kew, 1979). School leavers constitute another target group, anxious to express independence and individuality, but with limited incomes, low mobility and the loss of the ready-made recreation provisions and peer group support experienced at school (Hendry, 1981; Roberts, 1983).

People living in certain sorts of areas may also be constrained in their leisure, particularly those living in the inner cities, and some rural residents. In the inner city, physical decay in the form of poor and overcrowded housing and derelict land combines with the maximum concentration of social stresses: high density living, lower socio-economic groups, young people, people on low incomes and unemployed, single parent families, ethnic minorities and transient communities in which it is difficult to create social bonds and a sense of identity. Paradoxically, such areas tend to be closest to the greatest concentrations of commercial entertainments, such as cinemas, pubs, restaurants and discos, and often close to a wide range of public provisions such as sports centres, swimming pools and parks. However, accessibility is not just a matter of physical proximity, but also of social and financial accessibility too. Few inner city residents can afford city centre restaurant and cinema prices, and the nature of provision is often more geared to the tastes of tourists and the more affluent and mobile residents of the outer city, than to those of the immediate community.

Recreational deprivation can also be a problem in remote rural areas with dispersed and declining populations. Out-migration from such areas is selective, leaving behind the less qualified, poorer, older and less mobile sectors of the community; thus the social basis of deprivation, though smaller than in the inner city, may be no less acute. Furthermore, in the rural context, deprivation has a more obvious spatial dimension. The populations of many villages are insufficient even to support the most basic community facilities. Existing facilities, such as village halls and primary schools fall into disuse, and the residents who remain have greater distances to travel to use facilities in towns and few means of doing so as public transport services contract.

Urban and rural stereotypes, however, are being eroded, and are no longer a sound basis for planning. Approximately one-fifth of the population lives in rural areas, but this proportion is rising rapidly, especially in the south, as early retirees, commuters and high technology industries are more free floating, and no longer tied to towns and cities. This trend towards counter-urbanization is bringing to rural areas incomers who are affluent, mobile, articulate, and accustomed to urban standards of service provision. Measured against conventional social indicators, they are far from deprived, but they are moving to areas with little or no leisure provision. In many instances, there is a clash of cultures between the newcomers and the long established residents. Studies in Berkshire villages have found examples of newcomers establishing and running new clubs and committees, with resolute opposition from established villagers, who also lament their local pubs being 'up-marketed' into steak bars (Glyptis, 1987). Elsewhere, though, the opposite can be true: indigenous residents want to see their areas progress, and aspire to better facilities for their families, but meet strong resistance from new-comers who want to fossilize their new-found rural paradise. Furthermore, even residents in the remotest rural areas are no longer isolated and self contained communities: they have access to urban aspirations through the media. An urban way of life derives as much from awareness, attitudes and communications as from the basic fact of location.

Other types of areas can be equally short of provision. Many suburban council housing estates contain pockets of social deprivation; many private housing developments are completely without community recreation provision. Many tourist resorts are surprisingly short of recreation provisions attuned to the needs of local residents.

The concentration of non-participation among particular sorts of people and particular sorts of places has led agencies such as the Sports Council to undertake a clear shift of policy approach over the past 20 years. In the 1970s, the emphasis was firmly on providing facilities: in 1970 there were 27 sports halls in England; by 1980, there were nearly 800. By contrast, the 1980s were a decade of far greater community orientation, with much greater concern for overcoming the constraints faced by particular target groups. This was clear, for example, in new forms of grant-aid for 'Areas of Special Need', in special sports leadership schemes for unemployed people and inner city communities, and in schemes of peripatetic provision for rural areas. The Department of the Environment, through its Urban

Programme, has funded numerous inner city projects, thereby increasing local leisure opportunities and perhaps reclaiming derelict land and buildings too.

Many such schemes have achieved considerable success. Some have attracted large volumes of use and shown that latent demand for sport and physical activity has existed within the various target groups. Several, though, have been disappointing in their longer term impact. The Leicester sports leadership scheme for the unemployed, the STARS scheme of the early 1980s, was a case in point. The project consisted of a team of sports leaders working on an outreach basis, providing activity sessions at existing facilities throughout the city, and also going out into the community, into such venues as shopping centres and job centres, to get to know people and encourage them to come along and take part in activities. All activities were provided free of charge, and all were open to anyone, even though intended mainly for the unemployment. Over a three-year experimental period, with registered unemployed in Leicester at approximately 20 000, the scheme attracted 6400 users and over 30 000 attendances (Glyptis, Kay & Donkin, 1986). The project hit its target: 98% of those attending were unemployed people. Furthermore, the scheme succeeded in tapping latent demand: 40% of users had never previously taken part in the activities they did at STARS, and even among the other 60%, half had long since given up sport.

Once recruited or resurrected, however, participation was hard to sustain. For a committed minority (about one-tenth of the users) sport became a central interest: for quite some time they took part in the scheme at least once a week, some also took part in sport elsewhere, and many spent time with friends they met initially at the sports scheme; a few groups established their own self-supporting clubs. For the majority, however, sport was not to become a focal interest, but was simply something to taste or dabble in occasionally. Three-quarters of all STARS users attended five times or fewer. However, this is not necessarily a sign of failure, nor is it greatly at odds with the participation habits of young people generally. The scheme evidently provided a valued opportunity for casual 'dropping in'. Most unemployed people, it seems, are unlikely to become frequent sports players, but the fact that large volumes of use were sustained throughout a three-year period showed considerable demand for sports activities on an occasional basis.

Participation through the lifespan: the dynamics of participation

More recent studies have re-affirmed the difficulty of re-creating sustained participation by those whose involvement in sport has lapsed. Using recall methods, Roberts *et al.* (1990) studied patterns of participation throughout adulthood, concluding that (p. 16) 'The overwhelming likelihood was that anyone who withdrew from sport in early adulthood would be lost forever'. Whereas young people typically dabbled in a wide range of activities, some of them pursued only briefly or sporadically, with advancing years people's leisure became focussed increasingly around a smaller number of retained pastimes. Those who continued to play sport throughout the lifespan had begun with a wider repertoire of activities than those who had not, and they continued to draw upon that repertoire many years later. The key to long-term patterns of involvement or abstinence lay in the leisure activities taken up or given up in young adulthood. Beyond that phase, participants showed high levels of loyalty to sport: of those who had participated continuously up to the age of 25, 95% were still playing at age 30; of those playing continuously to age 30, 97% were still playing at 35. Policy-makers seeking to sustain participation into later life would therefore achieve more by preventing drop-out than by promoting take-up, i.e. by putting more effort into discouraging drop-out in young adulthood, rather than trying to rekindle lost interest among middle aged and elderly non-participants. As Roberts *et al.* stated (p. 19), 'many non participants had no intention of ever being tempted back into sport. They saw no circumstances which might persuade them to return. Some could not believe that anyone would want to persuade them'. Of all the sports ever played by those who had participated continuously since age 16, 73% had first been played before the age of 20. Young adulthood, of course, is also a critical phase when social class differences can be decisive, with middle class young people far more likely to take part in sport and recreation, and far more likely to have access to good facilities and information. However, the Roberts *et al.* study showed that members of the working class who did take part in sport at that stage were just as likely as their middle class counterparts to sustain their participation thereafter.

Lifestyles

Thus far this chapter has examined leisure activities, and especially sport, in isolation, rather than in the context of broader lifestyles. Roberts *et al.* produced evidence of broader sporting lifestyles among current users of indoor centres: of 1387 current participants in sport, 72% were taking part regularly in two or more different sports, 41% in three or more, and 21% in four or more. Among sporting unemployed people, Kay (1987) found that most were active in other ways too, involved in civic activities, hobbies, educational courses or private study. Using data from the *General Household Survey*, Gratton & Tice (1989) demonstrated a link between sports participation and subjective feelings of health and wellbeing. Asked whether, in the past year, they felt their health had been good, fairly good or not good, sports players were much more likely than non-participants to rate their health as good, and this was all the more true of sports players in the older age-groups and lower income bands. Sports participants were less likely than non-participants to smoke, and less likely to be heavy smokers; however, participants were more likely to drink, and more likely to be moderate or heavy drinkers. Sports participants also led more varied lifestyles: on average, sports participants in all age and income groups had taken part in nearly three times as many different leisure activities in the previous four weeks as non-participants.

Leisure lifestyle studies have tended to focus on sports participants rather than random samples of the population. Leisure lifestyles, in any case, exist within the broader context of lifestyles as a whole, including paid work, domestic work, chores and other activities. A study undertaken in Nottingham exemplified the many different ways that leisure fits in (Glyptis, McInnes & Patmore, 1987). The findings are based on interviews and time-budget diaries completed by 460 adults. The diaries required people to record all their activities of 15 minutes' duration or more over a three-day period, and for each activity to indicate its location within or out of the home, who they were with, and their perception of the activity; they were asked to describe it as leisure, work, duty, chore, personal care, or any other label they may care to use.

The diaries showed considerable variety of lifestyles. The sample as a whole recorded 119 different activities. Individual lifestyles, however, were much more specialized, the average diarist recording 48 events, made up of 15 different activities, 11 home based and 4 out of home. Women's activity patterns were more fragmented than men's and more home based. Men, on average, recorded 43 events, 35 of them home based. Women, on average, recorded 51 events, 45 of them home based. Home-based activities dominated the lifestyles of all social groups: 86% of all events took place there. Furthermore, much out-of-home activity took place in close proximity; over a third of out-of-home events took place within a mile of home. Overall, one third of all events (and 36% of all time) were recorded as leisure.

The diary records, in effect, form a personal biography for each individual, albeit for a brief episode in the lifespan. The archetypal single person led a fairly solitary existence, but single men, in particular, were more likely than couples or families to spend leisure time with friends and out of the home. One particular young man exemplified this pattern (Figure 16.1). On Sunday he slept in until a visitor arrived around noon, and then had an early afternoon breakfast, watched television and read the papers, and then went out to a sports fair with his girlfriend. He then went

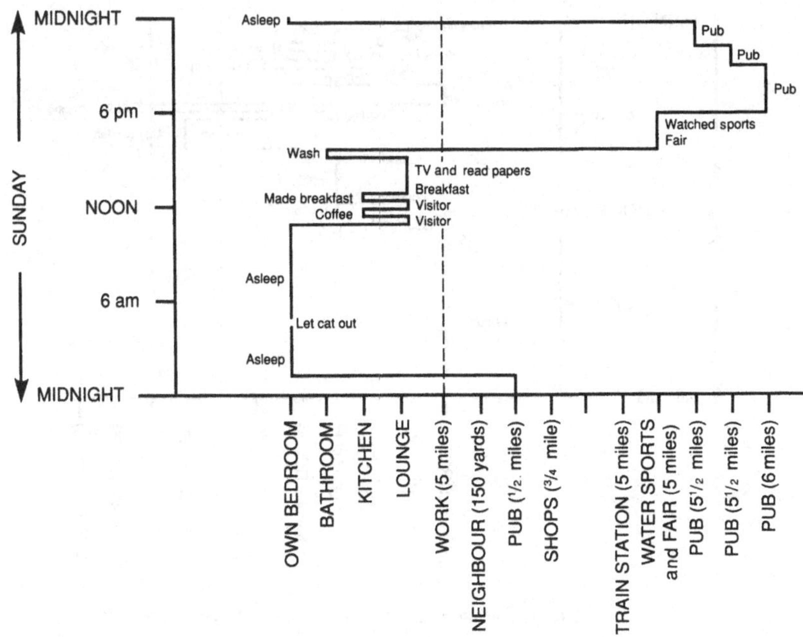

Figure 16.1. The flexible activity pattern of an illustrative single man.

to three pubs in quick succession, returning home towards midnight. He had complete freedom and flexibility to use his time as he pleased, and used that freedom to the full.

Another single man, however, illustrated how some single people spurn that flexibility, and construct for themselves a highly structured routine, indeed, so structured in this particular case that he might be labelled 'Clockwork Man' (Figure 16.2). He left the home only to go to work and go shopping, and apart from talking briefly to a neighbour while tidying the garden on Saturday, had no contact with other people throughout the entire diary period. His use of time was highly regularized, rooted in a commitment to exercise and meditation, with which he began each morning and evening. Music was another focus, with evening music practice on the first two days, and listening to the radio and records weaving an intermittent thread throughout. He had pervasive leisure interests, and a daily routine built around them.

Those patterns contrast sharply with that of a young mother, with a routine of unrelenting domesticity where leisure barely existed (Figure 16.3). Her lifestyle was dominated by their three young children, but a leisure theme also came through in the form of a love of gardening,

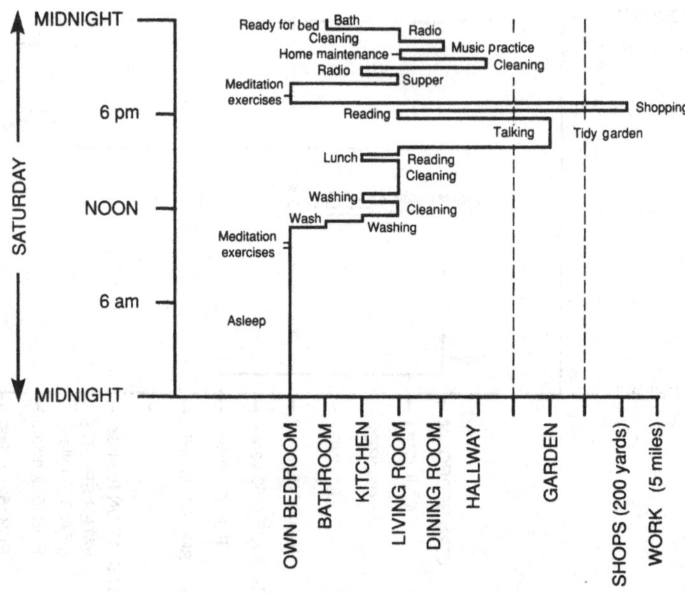

Figure 16.2. The routinized activity pattern of an illustrative single man.

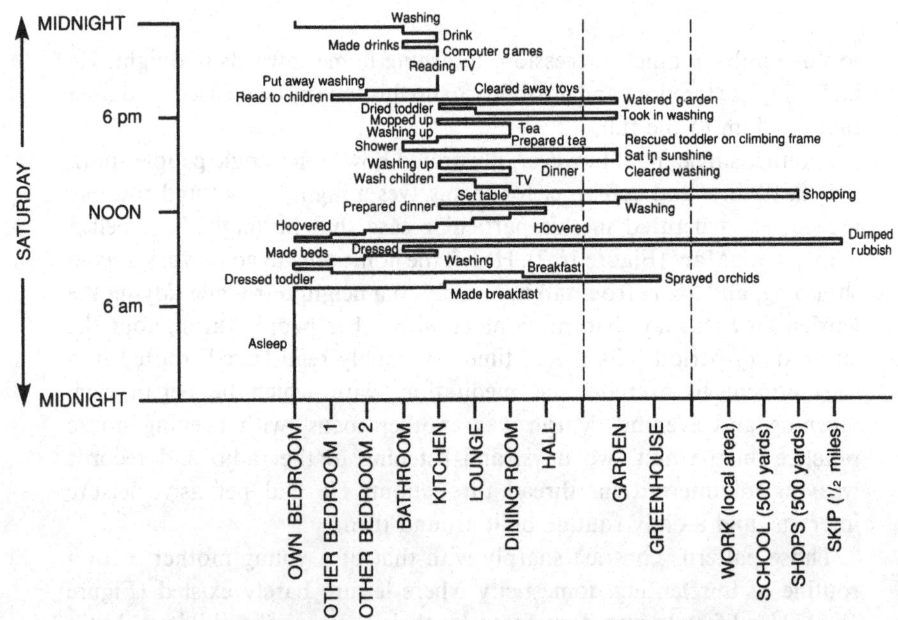

Figure 16.3. The fragmented activity pattern of an illustrative young mother.

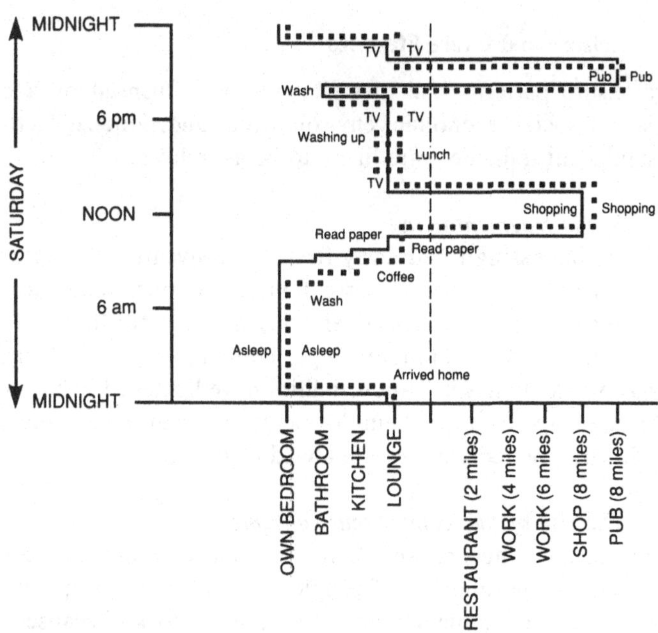

Figure 16.4. The symmetrical couple.

particularly cultivating orchids. Her routine was one of continual child-centred concern, punctuated with occasional release to the respite of the orchids. Her life was heavily home based and highly spontaneous; she was continually 'on call', attending to the toddler during the night and rescuing him from the climbing frame on Saturday afternoon! Leisure was a rare occurrence, but the gardening theme came through strongly in the structuring, if not in the consumption, of time, notably the mid-morning spraying of the orchids, and evening watering and potting sessions.

Leisure, then, took several different forms and differing degrees of prominence in these three lifestyles. It varied too with the degree of convergence or divergence of activities between different household members, ranging from the symmetrical couple (Figure 16.4), barely separable except for their different working hours, to asymmetrical couples, with both partners frantically busy, but hardly seeing each other. These different lifestyles differed not only in content but also in their social basis.

Leisure and future lifestyles

Future leisure patterns and lifestyles will be influenced by a complex interplay of social, economic, environmental and political factors. The most significant influences are likely to be as follows.

Social fragmentation

There is an increasing trend away from the conventional nuclear family towards smaller and more diverse households, including more single person households, more pensioner households, more 'dinkies' (double income with no kids), and more 'empty nesters', couples in their 50s and early 60s whose domestic responsibilities have lightened; who are beginning to disengage from paid employment; and who have many years of physically and mentally active life ahead of them.

The rise of the 'connoisseur consumer'

With increasing affluence and fewer social and family ties comes the emergence of large numbers of people who can invest in quality leisure experiences and equipment, and who want to do so because they are dedicated to particular interests, or because their leisure styles and equipment lend social status and exclusivity. This will mean increased consumption of non-utility goods, such as special cars, gourmet food, and stylish clothing, and increased demand for quality provision and customer care at leisure venues. This becomes all the more important at times of economic growth: more people are drawn into employment; people work longer hours with more overtime; hence leisure becomes more precious, and there is more money to spend per hour of leisure time. People want, and can afford, money-intensive leisure.

Social polarization

An increasing divide between rich and poor, the connoisseur consumers as compared with low income groups such as elderly people on state pension, single parents, unemployed people, and housewives with no income of their own. Leisure for these latter groups is likely to become even more home centred, more media-oriented, and less varied.

The role of the home as a leisure centre

The home is the best used leisure centre of all, but even in this context there are the 'haves' and the 'have nots'. The 'haves' have their own homes, with sufficient rooms and outdoor space to accommodate a variety of leisure functions, comfortably lit and heated, and well endowed with specialist leisure equipment and labour-saving devices. The 'have nots' live in crowded conditions where people and activities constantly compete for space, and where standards of servicing and comfort may make leisure activities impossible and irrelevant. The home is important as both a private and a social space. On the one hand, it offers a controlled autonomous space, over which others have few, if any, rights. It fosters a privatization of lifestyles. On the other hand, for the 'haves', it is an important focus of leisure and an arena for sociability. The home can be a leisure focus in its own right, continually changed to meet the changing tastes and requirements of the household. Increasingly, people are concerned with the appearance, ambience and style of their homes, and part of the enjoyment of that style is sharing it with others. The home is thus a venue for both self-centred and social leisure.

Increasing concern for health and environmental issues
Leisure tastes are influenced by the increased public awareness of environmental issues, conservation needs, the collapse of community cohesion, and the benefits of healthy lifestyles. There is increasing commitment to self development and self-enhancement, to community co-operation and self-help, to voluntarism, to prudent use of energy resources, and to sustainable forms of leisure and tourism.

Leisure as an important component of lifestyle, and it is likely to become increasingly important. Part of its meaning, however, derives from its counterbalance with work and with other aspects of lifestyle. Suggestions that leisure might soon become a substitute for work, however, appear to be unfounded. Studies of unemployed lifestyles, for example, make it clear that leisure, on its own, is rarely fulfilment enough; for most people its true place is appreciated only alongside other, purposeful activities (Glyptis, 1989). However, leisure is not merely an appendage of lifestyle that fits around everything else. For most people it is a positive contribution to lifestyle and the quality of life, helping to shape daily routines, and helping to construct and affirm their social position.

References

Boothby, J., Tungatt, M.F., Townsend, A.R. & Collins, M.F. (1981). *A Sporting Chance?* London: Sports Council.

Brightbill, C.K. (1963). *The Challenge of Leisure*, Englewood Cliffs, New Jersey: Prentice Hall.

Countryside Recreation Research Advisory Group (1970). *Countryside Recreation Glossary*, London: Countryside Commission.

Countryside Recreation Research Advisory Group (1980). *Digest of Countryside*

Recreation Statistics 1979, Cheltenham: Countryside Commission, Countryside Commission for Scotland & Sports Council.
de Grazia, S. (1962). *Of Time, Work and Leisure*, New York: Doubleday.
Dumazedier, J. (1967). *Towards a Society of Leisure*, New York: Free Press.
Fryer, D. & Ullah, P. (1987). *Unemployed People*. Milton Keynes & Philadelphia: Open University Press.
Glyptis, S. (1987). *Recreation in Expanding Residential Areas. A Study of West Berkshire and Theale*, Report to the Sports Council and Newbury District Council.
Glyptis, S. (1989). *Leisure and Unemployment*, Milton Keynes & Philadelphia: Open University Press.
Glyptis, S., Kay, T.A. & Donkin, D. (1986). *Sport and the Unemployed. Report on the Monitoring of Sports Council Schemes in Leicester, Derwentside and Hockley Port, 1981–1984*, London: Sports Council.
Glyptis, S., McInnes, H. & Patmore, J.A. (1987). *Leisure and the Home*, London: Sports Council & Economic and Social Research Council.
Gratton, C. & Tice, A. (1989). Sports participation and health. *Leisure Studies*, **8**, 77–92.
Green, E., Hebron, S. & Woodward, D. (1987). *Leisure and Gender. A Study of Sheffield Women's Leisure Experiences*, London: Sports Council & Economic and Social Research Council.
Hendry, L.B. (1981). *Adolescents and Leisure*, London: Sports Council and Social Science Research Council.
Kay, T.A. (1987). Leisure in the lifestyles of unemployed people: a case study in Leicester. Unpublished PhD thesis. Loughborough University of Technology.
Kelly, J.R. (1982). *Leisure*, Englewood Cliffs, New Jersey: Prentice Hall.
Kelvin, P. (1979). *A Memorandum on Leisure*, London: Sports Council & Social Science Research Council Joint Panel on Leisure and Recreation Research, Mimeo.
Kelvin, P. & Jarrett, J. (1985). *Unemployment: Its Social Psychological Effects*, Cambridge: Cambridge University Press.
Kew, S. (1979). *Ethnic Groups and Leisure*, London: Sports Council & Social Science Research Council.
Long, J.A. & Wimbush, E. (1985). *Continuity and Change: Leisure around Retirement*, London: Sports Council & Economic and Social Research Council.
Neulinger, J. (1974). *The Psychology of Leisure*, Springfield, Illinois: Charles C. Thomas.
Roberts, K. (1978). *Contemporary Society and the Growth of Leisure*, London: Longman.
Roberts, K. (1983). *Youth and Leisure*, London: Allen & Unwin.
Roberts, K., Minten, J.H., Chadwick, C., Lamb, K.L. & Brodie, D.A. (1990). *Sporting lives: a case study of leisure careers*. Paper presented to Canadian Congress of Leisure Research, Waterloo, Ontario.
Rodgers, H.B. (1977). *Rationalising Sports Policies. Sport in its Social Context*, Strasbourg: Council of Europe.
Social Trends 1990, London: Central Statistical Office.
Sports Council (1986). *A Digest of Sports Statistics for the U.K.* Information Series No. 7, London: Sports Council.

Sports Council (1988). *Sport in the Community: into the 90s*, London: Sports Council.

Talbot, M. (1979). *Women and Leisure*, London: Sports Council & Social Science Research Council.

Tokarski, W., Filipcova, B. & Glyptis, S. (eds) (1990). *Lifestyles. Theories, Concepts, Methods and Results of Lifestyle Research in International Perspective*, Prague: Czechoslovak Academy of Sciences.

Index